The Ecology of Snow and Ice Environments

Johanna Laybourn-Parry
Bristol Glaciology Centre, School of Geographical Sciences, University of Bristol, UK

Martyn Tranter
Bristol Glaciology Centre, School of Geographical Sciences, University of Bristol, UK

and

Andrew J. Hodson
Department of Geography, University of Sheffield, UK

OXFORD
UNIVERSITY PRESS

OXFORD

UNIVERSITY PRESS

Great Clarendon Street, Oxford OX2 6DP

Oxford University Press is a department of the University of Oxford.
It furthers the University's objective of excellence in research, scholarship,
and education by publishing worldwide in

Oxford New York

Auckland Cape Town Dar es Salaam Hong Kong Karachi
Kuala Lumpur Madrid Melbourne Mexico City Nairobi
New Delhi Shanghai Taipei Toronto

With offices in

Argentina Austria Brazil Chile Czech Republic France Greece
Guatemala Hungary Italy Japan Poland Portugal Singapore
South Korea Switzerland Thailand Turkey Ukraine Vietnam

Oxford is a registered trade mark of Oxford University Press
in the UK and in certain other countries

Published in the United States
by Oxford University Press Inc., New York

British Library Cataloguing in Publication Data
Data available

Library of Congress Cataloging in Publication Data
Library of Congress Control Number: 2011941968

Typeset by SPI Publisher Services, Pondicherry, India
Printed and bound by
CPI Group (UK) Ltd., Croydon, CR0 4YY

ISBN 978–0–19–958307–2 (Hbk.)
      978–0–19–958308–9 (Pbk.)

1006612004

1 3 5 7 9 10 8 6 4 2

Preface

'As far as we could see, the inland ice was an unbroken plateau with no natural landmarks. From the hinterland, in a vast solid stream, the ice flowed with heavily crevassed downfalls near the coast'.

Douglas Mawson *The Home of the Blizzard* (1915)

'Unseen and untrodden under their spotless mantle of ice the rigid polar regions slept the profound sleep of death from the earliest dawn of time. Wrapped in his white shroud, the mighty giant stretched his clammy ice-limbs abroad, and dreamed his age-long dreams'.

Fridtjof Nansen *Farthest North* (1898)

Ice and snow environments in the polar and alpine regions have always been a focus of fascination for humankind. There is something almost seductive about the stark beauty of icy places. They demanded exploration, and as the history books show extraordinary people like Nansen, Amundsen, Scott, Mawson, and Shackleton sought to reveal their secrets, in some cases at enormous human cost. The age of exploration paved the way for the large-scale scientific investigations of today, supported by the type of logistics and technology that the early explorers could never have imagined. The polar regions are warming more rapidly than lower latitudes and this has heightened interest in their ice sheets, glaciers, sea ice, and ice-covered lakes. Climate change and its possible impact on ecosystem function and biodiversity have become major scientific issues.

Ice sheets and glaciers have not traditionally been the domain of biologists, rather they have been the territory of physical and chemical glaciological investigations. It has only been in the last decade or so that glacial environments have come to be regarded as another ecosystem in the cryosphere. It has become evident that biological processes play an important role in the geochemistry of these environments. Like other extreme environments, such as Antarctic lakes, they are systems dominated by microorganisms. Today, glaciology demands an interdisciplinary approach involving microbiologists, geochemists, glaciologists, and modellers. Now it is not uncommon to find university research groups involving all of these disciplines working together in an integrated manner. The Bristol Glaciology Centre exemplifies this approach. The literature is expanding exponentially, but one of the major challenges is to elucidate in much more detail what the microbial communities of the cryosphere contribute to biogeochemical processes, particularly in subglacial environments.

While there are many excellent books dealing with specific components of cryosphere biology, there is no single volume that covers a wide range of ice and snow environments. This volume is intended for both the specialist and non-specialist. Many students of glaciology have little biological knowledge, and likewise biologists may lack knowledge of the physical and chemical aspects of the cryosphere. The important issue of climate change is embedded within each chapter, rather than being treated separately under one heading. We have covered a wide spectrum of cryospheric environments (excluding the permafrost) including snow, supra- and subglacial and ice sheet environments, sea ice, and lake ice. The first chapter is intended as a background introduction to the more detailed treatment of each environment in the succeeding chapters. We have also included a chapter on astrobiology because many of the ice environments are regarded as an analogue for extraterrestrial life. Lastly, we consider future directions, as some exciting technologies are being developed that provide much greater scope for studying life in glacial environments on Earth and on other planets.

We would like to thank our editors Ian Sherman and Helen Eaton at the Oxford University Press and the other staff involved in the production of this book. We are indebted to colleagues worldwide who have generously provided photographic illustrations and Paul Coles at Sheffield University for producing the black and white illustrations. M. Tranter acknowledges a Leverhulme Fellowship (R/4/RFG/2010/06000) during which this book was produced.

Johanna Laybourn-Parry, Bristol Glaciology Centre, University of Bristol

Martyn Tranter, Bristol Glaciology Centre, University of Bristol

Andrew Hodson, Department of Geography, University of Sheffield
May 2011

Contents

Abbreviations   ix

1. An introduction to ice environments and their biology   1

  1.1 Introduction   1
  1.2 Introduction to functional dynamics and the organisms   2
    1.2.1 Community structure and function   2
    1.2.2 Organisms   6
  1.3 The cryosphere: past and present   8
    1.3.1 The last glacial maximum   8
    1.3.2 Contemporary fluctuations of glaciers and ice sheets   8
    1.3.3 Snowball Earth   9
  1.4 Sea ice   12
    1.4.1 Nature of sea ice   12
    1.4.2 Sea ice communities   14
  1.5 Lake ice   16
  1.6 Glaciers   19
    1.6.1 Ice mass balance zones in glacial ecosystems   19
    1.6.2 Hydrological zonation in surface ecosystems   21
    1.6.3 Supraglacial lakes   23
    1.6.4 Water distribution in subsurface ecosystems   24
    1.6.5 Water in subglacial habitats   25
    1.6.6 Overview: broad structure and characteristics of glacial ecosystems   27
    1.6.7 Life on glaciers   30
  1.7 Snow   31
    1.7.1 Physical and chemical characteristics   31
    1.7.2 Biological activity in snow   35

2. Snow   37
  2.1 Snow as an environment   37
  2.2 Life on and in snow   37
    2.2.1 Snow algae   37
    2.2.2 Bacteria in snow   39
  2.3 Impact of snow on environments it covers seasonally   41
    2.3.1 Activity under the snow   41
    2.3.2 Impact of the release of accumulated nutrients in the snow pack at spring melt   44
    2.3.3 Variations in snow depth   46

3. Ice surface environments   **47**

3.1 Ice shelves   47

   3.1.1 Introduction   47

   3.1.2 Biology of ice shelf lakes   48

3.2 Glaciers and ice sheets   52

   3.2.1 Supraglacial habitats   52

   3.2.2 Spatial variations in the biota in supraglacial habitats   56

   3.2.3 Cryoconite   57

   3.2.4 Carbon cycling and biological production   64

   3.2.5 Other debris habitats, including the ice margin   71

4. Sea and lake ice   **73**

4.1 Sea ice   73

   4.1.1 Introduction   73

   4.1.2 Adaptations   77

   4.1.3 Community structure and production   78

4.2 Lake ice   92

   4.2.1 Introduction   92

   4.2.2 Community structure and production   95

5. Subglacial environments   **101**

5.1 Introduction   101

5.2 Biology of subglacial environments   104

   5.2.1 Wet-based glaciers   105

   5.2.2 Blood Falls   107

5.3 Life in glacial ice   110

5.4 Subglacial lakes   111

5.5 Lake Vida   117

6. Astrobiology   **121**

6.1 Introduction   121

6.2 Extraterrestrial cryospheric environments   122

   6.2.1 Mars   122

   6.2.2 Europa: a Jovian moon   125

   6.2.3 Enceladus: a small Saturnian moon   126

   6.2.4 Titan: a large Saturnian moon   127

6.3 Weaknesses of terrestrial analogues for extraterrestrial cryospheric environments   128

7. Future directions   **129**

7.1 Introduction   129

7.2 Priority field sites for future research   130

7.3 Remote sensing development   133

7.4 Sensor technology   135

7.5 Modelling   139

7.6 Molecular biology   140

7.7 Elucidating the evolution of extremophile communities   141

Glossary   **143**

References   **149**

Index   **173**

Abbreviations

AARs	accumulation area ratios
AMP	adenosine monophosphate
APR	adenosine 5´ phosphosulphate reductase
APS	adenosine 5´ phosphosulphate
ARA	acetylene reduction activity
BGE	bacterial growth efficiency
BIAs	blue ice areas
BP	bacterial production
BR	bacterial respiration
CF_x	concentration factor
CFUs	colony forming units
CO_2	carbon dioxide
cps	capsular polysaccharides
DAPI	4´,6-diamidino-2-phenylindole
DGGE	denatured gradient gel electrophoresis
DIC	dissolved inorganic carbon
DMS	dimethyl sulphide
DMSP	dimethysulfonioproprionate
DNA	deoxyribonucleic acid
DOC	dissolved organic carbon
DOM	dissolved organic matter
DON	dissolved organic nitrogen
E_k	shade adaptation indices
EPs	exopolymer particles
eps	extracellular polysaccharides
EPS	extracellular polymer substances
FISH	fluorescence *in situ* hybridization
FLBs	fluorescently labelled bacteria
GISP	Greenland Ice Sheet Project
GPR	ground penetrating radar
HGT	horizontal gene transfer
HNAN	heterotrophic nanoflagellates
HPLC	high performance liquid chromatography
IPCC	Intergovernmental Panel on Climate
LC-MS/MS	liquid chromatography mass spectrometry
LGM	Last Glacial Maximum
LIA	Little Ice Age

LIFE	laser-induced fluorescence emission
LIMCOs	lake ice microbial communities
LTER	Long Term Ecosystem Research Program
MSL	Mars Science Laboratory
NAO	North Atlantic Oscillation
NASA	National Aeronautics and Space Administration
NMR	nuclear magnetic resonance
OTUs	operational taxonomic units
PAHs	polycyclic aromatic hydrocarbons
PAR	photosynthetically active radiation
PCR	polymerase chain reaction
PKSs	polyketide synthases
PMF	peptide mass fingerprinting
PNAN	phototrophic nanoflagellates
POC	particulate organic carbon
PS	photosystem
PUFAs	polyunsaturated fatty acids
rRNA	ribosomal RNA
RUBISCO	ribulose-1,5-bisphosphate carboxylase/oxygenase
SAM	Sample Analysis at Mars instrument suite
SAR	Synthetic Aperture Radar
SPOT	Satellite Probatoire d'Observation de la Terre
TGM	temperature gradient metamorphism
UAVs	unmanned airborne vehicles
UUVs	unmanned underwater vehicles
UV	ultraviolet
VBRs	virus to bacterial ratios
w.e.	water equivalent
WISSARD	Whillans Ice Stream Subglacial Access Research Drilling

CHAPTER 1

An introduction to ice environments and their biology

1.1 Introduction

The cryosphere is a term derived from the Greek word *cryo* meaning cold. The term is used to describe that part of the world where water is frozen, either for long periods or seasonally. The cryosphere includes ice caps and ice sheets, glaciers, snow cover, and the sea and lake ice covers. It also includes frozen soils or permafrost. The cryosphere plays an important role in the Earth's climate system. Moreover, the vast polar ice caps, particularly in Antarctica, are effectively a history book. Ice contains gas from the atmosphere at the time the ice was formed thousands of years ago. Analysis of ice cores has allowed us to follow historical changes in carbon dioxide for example. On average the polar regions are experiencing greater climate warming than lower latitudes and this has resulted in a greater focus on the dynamics of their hydrology. The retreat of glaciers and the melting of ice caps and perennial sea ice over the North Pole are raising worldwide concern about increasing sea levels. Of particular concern is the rapid demise of highly reflective sea ice over the Arctic Ocean. Snow-covered sea ice has a reflectance (albedo) of 80%, whilst open water has a reflectance of just 6%. Even the albedo of snow-free sea ice is far greater than the open ocean (typically 50%). Therefore the radiation and heat budget of the Arctic will change markedly if complete sea ice removal occurs, because the amount of solar radiation absorbed by the sea surface will increase by 44% or more in certain places during the summer. Sea ice also insulates the upper ocean and restricts mixing by surface winds. The impacts of its loss are therefore rapid and not restricted to the marine environment, because there are signs that summer glacier melt-

ing and terrestrial ecosystem activity are also influenced by this sea ice amplification of climate change (ACIA 2005).

The greatest volume of ice is found in Antarctica, which contains 75% of the world's freshwater. Each year vast extents of the oceans in the northern and southern polar regions develop ice cover. Lakes in polar regions are covered by seasonal ice cover and in the highest latitudes by perennial ice covers, for example the lakes of the Antarctic Dry Valleys. Glaciers are a major feature of the polar and many alpine regions. At first glance these extreme environments do not appear to offer habitats that can support life. The sea ice has long been recognized as an environment that teems with life, particularly in the spring and early summer before it breaks up. Glaciers, lake ice, and snow have only really attracted attention recently and are proving to be exciting sites for study. Contrary to the notion that glaciers are frozen rivers devoid of life we are now coming to the view that they are another important ecosystem in the cryosphere (Hodson *et al.* 2008).

This introductory chapter is intended to introduce the reader to the basics of the nature of ice environments and their biota. Biologists may have only a passing acquaintance with glaciers, snow, and lake and sea ice, while earth scientists may have little knowledge of the communities that inhabit them. Subsequent chapters will develop each type of habitat in detail.

The discovery of numerous lakes under the Antarctic Ice Sheet has created great excitement as to what they may possess in terms of communities. Many of the lakes may be interconnected by subglacial streams and rivers, creating a complex drainage system. It has been estimated that the volume of Antarctic subglacial lakes probably exceeds

$10\,000\,km^3$ (Dowdeswell and Siegert 1999). Lake Vostok at $5400\,km^3$ and Lake $90°E$ ($1800\,km^3$) are the largest. At present we have only limited indications of what type of life they harbour based on an ice core from above Lake Vostok.

Glaciers have not traditionally been regarded as aquatic ecosystems. However, extensive research on surface Antarctic lakes has shown that provided there is liquid water, even in the most extreme environments, communities of functional organisms exist. The surfaces of glaciers are subject to significant melt in summer providing a life-supporting environment. We now know that simple microbial communities function in glacier habitats and we are gradually gaining a picture of their contribution to important biogeochemical processes such as carbon and nitrogen cycling. The bases of glaciers and ice sheets are other sites where liquid water may occur and life is supported.

No book covering life in ice would be complete without a consideration of sea ice. This is an environment on which there is very substantial literature. However, it is useful to draw comparisons between the sea ice environment, which is productive and has a high biodiversity including significant populations of metazoans, and the more unproductive ice environments. Lake ice has been ignored as a potential habitat until fairly recently, but it too supports life. Lake ice covers are quite diverse globally and differ depending on whether they are perennial or annual and on the extent of snow cover.

Red snow seen in the polar regions and in alpine locations has been noted from ancient times. The advent of microscopy showed that it was caused by 'algae', commonly known as snow algae. More recent research has demonstrated that snow, even in the most harsh of environment of the South Pole, supports bacterial life (Carpenter *et al.* 2000).

1.2 Introduction to functional dynamics and the organisms

1.2.1 Community structure and function

Microorganisms dominate the communities of ice and snow environments. Higher organisms (Metazoa) are usually sparse or lacking. This dominance of Bacteria, Algae, and Protozoa is a charac-

teristic exemplified by the Antarctic continent, which is a microbiological domain. The penguins and seals that are frequently seen on the continent's margins derive their food from the sea and are part of the marine ecosystem. They nest, or in the case of seals bask and pup, on the land, but derive no other support from it. As a general rule, as environments become more challenging and extreme, their food webs become more truncated. Extreme aquatic environments are characterized by simple truncated food webs whose components are microbial.

Life is only functional where there is liquid water, even if it is present only as a thin film around mineral particles. Effectively, ice and snow environments are aquatic and their microbial communities have much in common with 'traditional' aquatic ecosystems. In order to understand the functional ecology of extreme cold environments we need to explore how life functions in 'traditional' aquatic ecosystems.

Prior to the 1980s, studies on aquatic environments (lakes, the sea) were predicated on the view that bacteria and very small protozoans were of no consequence in the general scheme of aquatic ecosystem function. The procedure for collecting zooplankton and phytoplankton samples was to use nets with different-sized meshes that did not retain bacteria and small protozoa. Then an important change occurred and whole water samples were examined. At this time fluorescent dyes became available, like 4′,6-diamidino-2-phenylindole (DAPI) and acridine orange that were very effective in staining bacteria so that they could be seen and enumerated by fluorescence microscopy. These dyes were also excellent for staining small flagellated protozoans. Where bacteria had been studied previously by growing bacteria from field samples on different types of agar plates, the calculated numbers were low. Once direct counts were made using fluorescence microscopy it became apparent that the plating procedure grossly underestimated the number of bacteria in the plankton. Moreover, it was clear that since numbers derived from plate counts were low compared to direct counts, only a small fraction, probably around 2%, of bacteria are culturable. The majority will not grow in the laboratory using current culture technology.

As a result of this new approach for investigating aquatic communities a new theory was postulated, that of the microbial loop (Pomeroy 1974; Azam *et al.* 1983) (Figure 1.1). At that time the data were limited; now we have a vast data base on the organisms and their functional ecology. What started out as a relatively simple model that revolutionized the way we viewed the functional ecology of lakes and the sea and subsequently other aquatic environments, is today highly complex as we have discovered much more about the biodiversity of microorganisms and their physiology and biochemistry. What we understand today is not a simple 'loop' operating in tandem with the classic food chain, but a complex food web that integrates microbial processes in the biogeochemical cycling of carbon and other elements (Figure 1.1). The sizes and the terminology applied to different size categories of aquatic organisms are shown in Table 1.1.

The microscopic organisms that collectively form the phytoplankton are a taxonomically diverse assemblage, including algae (e.g. diatoms, desmids),

Table 1.1 Size categories of microorganisms.

Constituent organisms	Size range
Viruses	retained by a 0.02 µm filter
Bacteria, coccoid Cyanobacteria	0.2–2.0 µm
Heterotrophic and autotrophic nanoflagellates, small ciliates, small naked amoebae, small dinoflagellates	2.0–20 µm
Ciliates, amoebae, dinoflagellates, diatoms, larger Cyanobacteria	20–200 µm

Organisms 0.2–2.0 µm often have the prefix 'pico' when found in the plankton (picoplankton), organisms 2.0–20 µm have the prefix 'nano', and larger organisms up to 200 µm have the prefix 'micro'. Note that some groups such as the ciliates and the dinoflagellates exhibit a wide size range among their species.

photosynthetic Protozoa commonly called phytoflagellates (e.g. dinoflagellates, chrysophytes, cryptophytes, euglenoids), and the Cyanobacteria. Thus the term phytoplankton is generic. Many of these organisms are characteristic of ice and snow environments. They carry out the process of photosynthesis, the first step in the carbon cycle. Photosynthesis is the process through which autotrophic organisms (plants) fix inorganic carbon (from carbon dioxide (CO_2)) combining it with water to produce new living material or biomass. Photosynthetic pigments harvest light energy within specific wavelengths of the visible light spectrum and this provides the energy to drive the reaction. In eukaryotic organisms photosynthetic pigments are contained in cellular organelles called chloroplasts. The process of photosynthesis can be summarized thus:

$$2H_2O + CO_2 + \text{light energy} \rightarrow (CH_2O) + H_2O + O_2$$

Autotrophs also require sources of major inorganic nutrients such as phosphorus and nitrogen, and trace nutrients such as iron and manganese to create new biomass. In many aquatic environments the major nutrients may decrease to limiting concentrations during an annual cycle and this constrains photosynthetic activity. Light is fundamental to driving photosynthesis, the first step in the transfer of carbon up the food chain. Photosynthetic organisms that live in aquatic

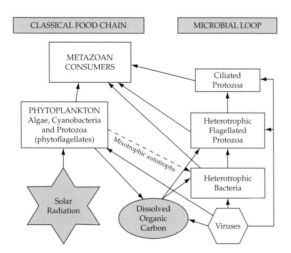

Figure 1.1 Structure of an aquatic pelagic food web. The concept of the microbial loop was postulated on the basis of limited existing data and it was tacked onto the classic food chain. Subsequent research revealed that the microbial 'loop' and the classic food chain form a complex interacting trophic web. In freshwater metazoan consumers are vertebrates (fish) and invertebrates (crustaceans, rotifers, insect larvae). Some of the phytoplankton are mixotrophic, feeding on bacteria as well as being autotrophic. A number of species are also capable of taking up dissolved organic carbon. Some heterotrophic flagellates exploit dissolved organic carbon as well as bacteria. Viruses infect the phytoplankton, bacteria, and heterotrophic protozoans. They lyse cells and rapidly recycle carbon to the pool.

environments effectively live in the shade. Light is attenuated as it enters a water column and follows an exponential decrease to the point where only 1% of total photosynthetically active radiation is present. This is called the compensation point, where carbon fixed during photosynthesis equals the carbon released during respiration. Thus there is no net gain of carbon to the organism. The phytoplankton provide a crucial energy source for crustaceans (e.g. krill, copepods) in the zooplankton.

During the process of photosynthesis, the phytoplankton release a proportion of their production (the photosynthate) into the environment as dissolved organic carbon (DOC). The pool of DOC provides an energy source to bacteria (Figure 1.1). The level of bacterial growth or production that can be achieved depends on the chemical composition of the DOC pool and molecular weight of its fractions, as well as the availability of inorganic nutrients such as nitrogen and phosphorus. Just like the autotrophs, bacteria require sources of nitrogen and phosphorus to support the production of new biomass. The conventional view is that the DOC pool accessible to heterotrophic bacteria is comprised of two distinct fractions. The first, a 'light fuel', consists of a wide array of organic compounds including dissolved free amino acids, dissolved free carbohydrates, and low molecular weight (<200 Da) organic acids, lipids, vitamins, hydrocarbons, polyphenols, and enzymes. This pool has been shown to be rapidly utilized and turned over by the heterotrophic bacterial community (Fuhrman 1987). However, this is only a small fraction of the total DOC pool (Amon and Benner 1996). The major portion of the DOC pool is composed of higher molecular weight material in the form of proteins, polysaccharides, and humic substances. This material is less energetically favourable because it requires hydrolysis by bacterial enzymes into monomers and oligomers before it can be assimilated (Chrost et al. 1989; Coffin 1989). Bacteria excrete these enzymes into the medium and then assimilate the products of hydrolysis. Nevertheless, there is evidence that bacteria can use this higher molecular weight faction effectively to sustain growth (Amon and Benner 1996; Meyer et al. 1997).

Bacteria are predated by complex communities of protozoans, including heterotrophic nanoflagellates,

some small ciliates, and some species of microscopic metazoans called rotifers. The major predators of the bacterial community are usually the heterotrophic nanoflagellates. Among the phytoflagellates there are some species that are able to feed heterotrophically on bacteria and/or DOC. This phenomenon of combining phototrophy with heterotrophy is termed mixotrophy (mixed nutrition) (see Figure 1.1). Some genera are well-known mixotrophs, for example *Dinobryon* (Figure 1.2), which can derive up to 50% of its carbon intake from feeding on bacteria. Another mixotrophic group that is particularly common in polar environments is that of the cryptophytes and, in some instances, for example in Antarctic Lake Fryxell, they can have a greater predatory impact on the bacteria than the heterotrophic nanoflagellates (Roberts and Laybourn-Parry 1999) (Figure 1.3). Why do these photosynthetic creatures feed on bacteria when they can survive by photosynthesis? They either do it to gain carbon when photosynthesis is inhibited by low levels of photosynthetically active radiation (PAR), as might be the case under thick, sediment-laden ice or during the polar winter, or to gain nutrients such as phosphorus and nitrogen when these elements are limiting in the surrounding water. By feeding on bacteria they can access organic nitrogen and phosphorus to support photosynthesis. Mixotrophy is an important survival mechanism in extreme environments giving those organisms that possess the capability a strong competitive advantage. The nanoflagellates are in turn prey to ciliated protozoans, rotifers, and small crustaceans in the zooplankton.

In recent years aquatic scientists have gained an appreciation of the important role played by viruses in the sea and freshwater (Figure 1.3D). As in the case of bacteria, the study of viruses in aquatic environments has been possible because of advances in the development of fluorochromes and fluorescence microscopy. The average abundance of free viruses in freshwaters is $4.1 \times 10^{10} \, L^{-1}$ and in the sea from 6.7×10^{7} to $1.7 \times 10^{10} \, L^{-1}$ (Maranger and Bird 1995). In polar waters the density is much lower, for example in oligotrophic freshwater lakes between 0.56×10^{9} and $0.74 \times 10^{9} \, L^{-1}$ (Säwström et al. 2008).

Viruses are not strictly living organisms because they cannot reproduce. In order to replicate they have to infect a host and turn its cells into virus-producing factories. Viruses are known to infect bacteria, protozoa, and algae in aquatic environments. The majority of viruses in the sea and freshwater are parasites of bacteria (bacteriophages) and they may have the effect of causing the destruction of bacterial cells (lysis). This short circuits the microbial loop by returning carbon to the pool before it can be consumed by heterotrophic protozoans and other organisms. In some environments the viral destruction of bacteria equals the impact of predation imposed by heterotrophic nanoflagellates (Weinbauer 2004). Viruses can also live within a host without causing its destruction, as happens in the lytic cycle, where they are passed by bacteria to their progeny during cell division. This is termed the lysogenic cycle during which the viral nucleic acid recombines with the host genome. Once the virus has inserted itself into the host chromosome it is called a prophage, and the cell harbouring the prophage is called a lysogen. The lysogenic cycle can continue until some factor, or factors, triggers the lytic cycle (Wommack and Colwell 2000). It is proposed that lysogens have an advantage over their non-lysogenic counterparts in oligotrophic unproductive environments where available hosts

will be limited. However, at present there are insufficient data on the relationship between lysogeny and the productivity of the environment to substantiate this argument (Weinbauer 2004).

The microbial elements within aquatic environments play a cardinal role in the biogeochemical cycling of carbon, nitrogen, and phosphorus. All heterotrophic organisms excrete phosphorus (mainly as orthophosphate) and nitrogen (mainly as ammonium) as a consequence of metabolizing assimilated energy from their food intake. As indicated above the photosynthetic elements of the community and bacteria need a source of inorganic nitrogen and phosphorus to grow. The recycling of nitrogen and phosphorus by protozoans is at times extremely important in sustaining the physiological functioning of the autotrophic and bacterial community.

As we have seen above viruses release carbon into the environment by the lysis of bacteria and other organisms like protozoans and algae, and this may be significant. For example, Fischer and Velimirov (2002) estimated that the viral lysis of bacterial cells in a productive (eutrophic) lake could potentially release 5–39 μg C L^{-1} day^{-1}, which corresponded to

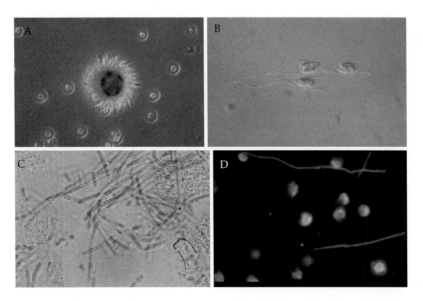

Figure 1.2 (A) *Mesodinium rubrum*: a ciliate fixed in Lugol's iodine, where the dark staining areas are the endosymbiotic cryptophyte; (B) *Dinobryon*: a colonial phytoflagellate (photo courtesy of W. Vincent); (C) Cyanobacteria *Leptolyngbya fragalis* (photo courtesy of Jeff Johansen and Mark Schneegurt (www.cyanosite.bio.purdue.edu)); (D) cryptophytes under epifluorescence microscopy. *M. rubrum*, *Dinobryon*, and cryptophytes are all mixotrophic species. See also Plate 1.

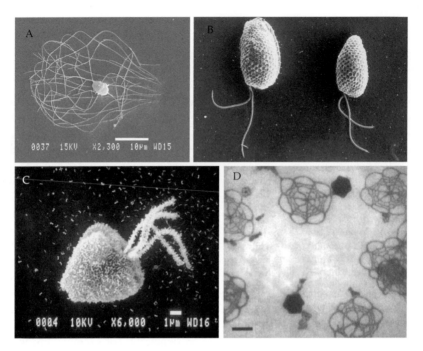

Figure 1.3   (A) Heterotrophic choanonanoflagellate *Diaphanoeca grandis*; (B) cryptophyte phytoflagllates; (C) *Pyramimonas* species; (D) viruses: dark pentagonal structures with scales from phytoflagellates. A–C courtesy of G. Nash, Australian Antarctic Division.

29–79% of bacterial production. In two very unproductive Antarctic lakes (Lake Druzhby and Crooked Lake) viral-induced bacterial mortality was higher, ranging from 38% to 251%, which contributed up to 69% of carbon supplied to the DOC pool (Säwström *et al.* 2007a). Other sources of DOC result from faecal outputs, decomposition, and allochthonous inputs (from outside the ecosystem). The latter can occur by aerial deposition from neighbouring ecosystems. For example, Arctic glaciers may receive aeolian carbon deposition from the surrounding tundra and can thus be described as subsidized ecosystems, at least partially. Estuaries and steams are classic examples of subsidized ecosystems receiving allochthonous carbon inputs from other ecosystems, in the case of estuaries from rivers and salt marshes and in streams from overhanging trees at the autumnal abscission.

1.2.2 The organisms

1.2.2.1 *Prokaryotes: Bacteria and Archaea*
Bacteria are described as prokaryotes, meaning that they lack a nucleus. Unlike eukaryotic organisms where the nucleic acid is contained in chromosomes within a nucleus bounded by a nuclear membrane, in bacteria the nuclear material appears as a central mass lacking a nuclear membrane. Bacteria were the first life form on the planet. The first forms were thought to be anaerobic, heterotrophic organisms. These were probably followed by what are known as anoxygenic bacteria that were capable of undertaking photosynthesis where oxygen was not produced. These were followed in the evolutionary sequence by the Cyanobacteria that undertake photosynthesis resulting in the production of oxygen. Other groups evolved such as the nitrifying and denitrifying bacteria and iron-oxidizing bacteria.

It should be noted that our understanding of the taxonomy of prokaryotes is still very limited. Part of this relates to the problem of culture. It is estimated that only around 2% of prokaryotes are culturable. Increasingly, a variety of molecular techniques have been applied to the study of aquatic bacteria. In particular, the potential of direct sequencing of the polymerase chain reaction (PCR) amplified 16S rRNA gene for phylogenetic comparison is well documented, and has been applied to the identification of new species, from, for example, Antarctic lakes, for nearly two decades (Dobson *et al.* 1993). The PCR-based detection technique has been further developed to use a denaturing gradient gel to separate PCR-amplified DNA fragments based on their denaturing (strand separation) characteristics and hence their DNA sequence. The particular power of this technological development in community studies is that it selectively amplifies predominant members of the community. By sequencing individual fragments, it is now possible to identify the dominant members of the community. The technology is evolving all the time, for example an alternative DNA sequence-based detection technique, which does not rely on the need for PCR amplification, is fluorescence *in situ* hybridization. This advance has been applied successfully to a number of Antarctic lake systems (Pearce 2003; Pearce *et al.* 2003, 2005) and Antarctic cryoconites (Foreman *et al.* 2007). The exciting advent of environmental metagenomics involving the genomic analysis of microorganisms by direct extraction and cloning of DNA from a whole assemblage of organisms (Handelsman 2004), allows questions to be asked about the physiological potential or function of a whole community.

There are two distinct groups of prokaryotes, the Archaea and the Bacteria. The Archaea are characteristic of extreme environments, being tolerant of high salinites and extreme temperatures. Many are methanogens (methane producing). Aerobic and anaerobic groups occur. They can be broadly divided into three groups: the methane-producing Methanoarchaea, the Haloarchaea, which are highly salt tolerant, and a group of extreme thermophiles. The Bacteria are a diverse group physiologically and include photosynthetic, chemoautotrophic, and heterotrophic bacteria. Some are facultative anaerobes, meaning they are capable of living in the absence of oxygen, while others are obligate anaerobes, and others are aerobic. Heterotrophic species are, as indicated above, very important in carbon cycling in aquatic environments. Among the photosynthetic species, the Cyanobacteria are very conspicuous and ubiquitous in aquatic environments worldwide (see Figure 1.2C). They include filamentous forms and unicells. Greater detail on these various groups can

be gleaned from standard microbiological textbooks, for example by Perry *et al.* (2002).

As we shall see in Chapters 3 and 4, the Cyanobacteria are a very important group in surface ice environments, particularly water bodies on ice shelves. They possess oxygenic plant-like photosynthesis and many species are capable of fixing atmospheric nitrogen, thereby enabling the successful colonization of environments where this essential element may become limiting to growth.

1.2.2.2 Protozoa

The Protozoa are eukaryotic microorganisms. Unlike the prokaryotes described above, they possess one or more nuclei bounded by nuclear membranes. The Protozoa used to be classified in the Animal Kingdom, but many of their members were also classified as Algae by botanists. This confusion was resolved when a new classification of the living world was proposed (Whittaker 1969; Margulis 1974). In the new system the living world was divided into five 'kingdoms': the Monera (prokaryotes), and the Protista, which includes the Protozoa, Plantae, Fungi, and Animalia. The Kingdom Protista is effectively an artificial group containing many organisms with no evolutionary affinities. The Sub-Kingdom Protozoa contains numerous phyla, some of which are entirely parasitic. There are two phyla that are relevant to aquatic and soil environments: the Phylum Sarcomastigophora and the Phylum Ciliophora. While the majority of their members are free-living, they do contain some important parasitic species, for example a number of heterotrophic flagellate species that cause diseases like sleeping sickness and leishmaniasis.

The Sarcomastigophora contains what are commonly called the amoebae, a complex group both morphologically and physiologically, and the flagellates. The flagellates typically have one or more whip-like structures called flagella that are used for effecting movement and aid in feeding. They fall into two distinct groups: the Phytomastigophora, which contains the photosynthetic flagellates (previously classified as Algae) (see Figure 1.3B, C), and the Zoomastigophora (see Figure 1.3A), which are heterotrophic, feeding mainly on bacteria. The flagellates, both photosynthetic and heterotrophic, are very important groups in aquatic environments.

However, the nutritional physiology of the phyto-flagellates is not necessarily clear cut. For example the dinoflagellates (Figure 1.4) are classified as phyto-flagellates, however only about half of the species contain photosynthetic pigments, the rest are colour-less and heterotrophic. Many of the species carrying photosynthetic pigments are mixotrophic, making them capable of both photosynthesis and heterotro-phy. Among other phytoflagellate groups mixotrophs are also common, for example the cryptophytes (see Figures 1.2 and 1.3), a common group in polar waters. Further information can be obtained from Leadbeater and Green (2000) and Scott and Marchant (2005).

The Phylum Ciliophora contains the ciliated protozo-ans and is one of the most uniform protozoan groups taxonomically. Ciliates are a conspicuous element of aquatic environments, in both bottom sediments and in the plankton (see Figure 1.2A). Some are large enough to be visible to the naked eye, particularly if they contain pigments, while others are only 5–10μm in length. Ciliates are typified by the possession of cilia, complex rows of hair-like structures that produce coordinated beating that effects movement and plays a role in feed-ing. Most ciliates possess a cell mouth or cytostome, which may have very complex ciliary structures associ-ated with it that enable filter feeding. Ciliates feed on flagellates, algae (e.g. diatoms) and bacteria, while some species are predatory on other ciliates. They are hetero-trophic, but some ciliates, notably the oligotrichs, are mixotrophic. They feed on phytoflagellates and steal or sequester the plastids (chlorophyll-containing organelles) of their prey. They then engage in photosyn-thesis using the sequestered plastids. Other ciliate spe-cies typically contain symbiotic algae called zoochlorellae. The relationship confers benefits to both species. The algae have access to phosphorus and nitrogen, the excre-tory products of the ciliates, as well as shelter within a highly motile organism that can position itself in an ideal light climate to support photosynthesis. The ciliate gains some of the products of photosynthesis to supple-ment its carbon budget. Further information on ciliates can be found in Foissner *et al.* (1999) and the chapter on ciliates in Scott and Marchant (2005).

1.2.2.3 Algae
These are effectively plants and include both micro-scopic and macroscopic species such as seaweeds. A particularly conspicuous group in some ice environ-ments is that of the diatoms (members of the Class Bacillariophyceae). They possess a pectinaceous cell wall that is impregnated by silica constructed in two distinct halves called valves (see Figure 1.4A, B). They may be solitary cells or form filamentous colonies. Because they need silica to construct their cell walls, the concentrations of silica in the environment can play an important role in limiting their growth, along with nitrogen and phosphorus. Along with the Cyanobacteria and protozoan phytoflagellates, the algae constitute what is known generically as the phytoplankton.

1.3 The cryosphere: past and present

1.3.1 The last glacial maximum

Several phases of ice sheet growth and decay occurred between 10000 and 120000 years ago during the most recent ice age (Siegert 2001). Growth to maximum ice limits (the so-called Last Glacial Maximum or LGM), as shown in Figure 1.5, was broadly synchronous throughout the cryosphere and was attained between 26500 and 19000 years ago (Clark *et al.* 2009). The tim-ing and full extent of the LGM is still subject to debate, especially in the case of those ice sheets with marine limits. However, the LGM ice extent was at least three times the present-day ice extent of about 16 million km² (Figure 1.5). The transition from LGM conditions to those conducive to deglaciation began 19000 to 20000 years ago following an increase in northern hemisphere insolation (e.g. Clark *et al.* 2009). The onset of deglaciation in Tibet and the southern hemi-sphere was later though, starting as recently as 14000 years. Although Patagonian and West Antarctic Ice Sheets were once sizeable, it was the northern hemi-sphere changes in ice extent that were by far the most significant. In fact several northern ice sheets have completely disappeared: the most notable being the Laurentide and Barents Ice Sheets. It therefore follows that these ice masses represent much of the 20% or more of the Earth's surface that has lost its icy glacial habitats over the last 20000 years.

1.3.2 Contemporary fluctuations of glaciers and ice sheets

Most glaciated regions of the Earth today are respond-ing to more recent climate changes that began with the end of the so-called Little Ice Age (LIA). This was a minor readvance that took place within the sixteenth and nineteenth centuries and left the most obvious evidence of glacial advance and retreat patterns visible today. In many cases, it was this readvance that led to the development of the moraines seen in the forefields of alpine glaciers today (Benn and Evans 2010). Glacier retreat from these moraines generally began in the late nineteenth and early twentieth centuries, producing recently deglaciated land surfaces composed of a mixture of supraglacial, englacial, and subglacial sediments (tills). The origins of these forefield sediments therefore means ecological succession and soil development following deglaciation are intricately linked to the glacial ecosystems once present above them.

Changes in glaciers and ice sheets since the LIA can to some extent be characterized by the instrumental records. Scientific studies of glaciers were initiated as early as the 1830s in the European Alps by Karl Schimper, Louis Agassiz, and colleagues. However, a network of observations of glacier mass balance (see Section 1.6.1) was not forthcoming until the 1950s. Recent assessments of glacier changes since this time have been presented by the Intergovernmental Panel on Climate (IPCC) (2007) and are shown in Figure 1.6. Of interest is the general decline in glacier covers, which has been sustained in most parts of the cryosphere, but has been offset in places by short-lived increases in winter snowfall. Thus parts of western Norway demonstrated minor readvances during the 1990s that led to deforestation, at least at the immediate ice margin. The incorporation of forest carbon into subglacial tills in this way again demonstrates the intricate link between glacial habitats and those of their forefields.

1.3.3 Snowball Earth

There were a series of global glaciations some 750–570 million years ago, each of which lasted for millions of years (Kennedy *et al.* 1998). These so-called Neoproterozoic glaciations gave rise to tills that are often associated with iron formations, which are capped with distinctive 'knife sharp' or 'cap' carbon-

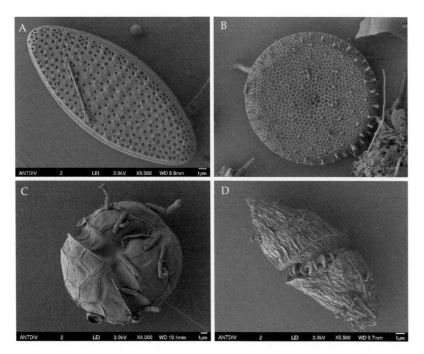

Figure 1.4   (A, B) Diatoms from Antarctic sea ice; (C, D) dinoflagellates from Antarctic sea ice. Photos courtesy of Harriet Paterson and the Australian Antarctic Division.

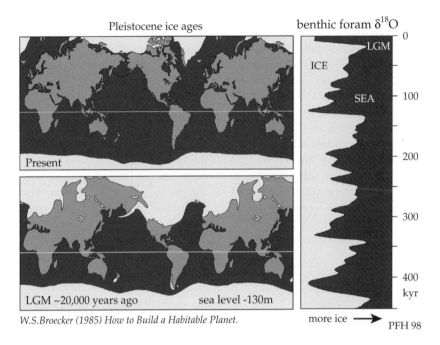

Pleistocene ice ages

benthic foram δ^{18}O

W.S.Broecker (1985) How to Build a Habitable Planet.

Figure 1.5 A comparison of the extent of permanent glaciers and sea ice at the present day (top left panel) compared with the last glacial maximum (LGM) (bottom left panel) some 20 000 years ago (20kyr). The oxygen isotope composition found in the carbonate shells of small organisms that used to live on the sea bed (so-called benthic forams) records the growth and decay of the great ice sheets over North America and northern Europe over the last 500 000 years (part of the Pleistocene epoch, which extends from ~2.6 million to ~12 000 years ago). From Broecker (1985), with permission of Wally Broecker and Eldigio Press.

ates (Schrag and Hoffman 2001). These carbonates, which have unusual textures, were deposited in deep seawater, along with other direct precipitates from seawater, and are characteristic of having been deposited directly from warm seawater. They also have very light δ^{13}C (Hoffman *et al.* 1998). Biological activity in the oceans locks up light carbon (^{12}C) into organic matter in preference to heavy carbon (^{13}C), which remains in the ocean water driving the ratio of the isotopes, δ^{13}C, towards positive or heavy values. So carbonates precipitating from seawater when biological productivity is high have heavy values of δ^{13}C, and have light values when biological productivity is low. The combination of cold climate tills overlain by deposits from a warm, deep ocean led to the formation of the 'Snowball Earth' hypothesis, which states that worldwide glaciation extended from the poles through the equator, and that marine biological productivity was very low (Figure 1.7). This probably arose because of the particular alignment of the continents and the variation in the receipt

of solar radiation, such that there was periodic runaway glaciation from the poles to the equator (Hyde *et al.* 2000). Debate is still ongoing about whether the ice at the equator was relatively thick, of the order of hundreds of metres (Hoffman *et al.* 1998), or relatively thin, of the order of metres (McKay 2000), because of the effect that a hard, frozen ocean would have on the survival of marine metazoans, which expanded greatly in magnitude and diversity after Snowball Earth had thawed (Hoffman *et al.* 1998). Indeed, some authors favour a 'Slushball Earth', with some areas of the equatorial ocean staying ice-free, so allowing marine metazoans to freely exist and to diversify during the cold bottleneck, then to expand their range and further evolve in uncolonized areas of the ocean during deglaciation (Hyde *et al.* 2000). Mean annual equatorial temperatures in the range of approximately –50 to –30°C are required for the persistence of a Snowball Earth (McKay 2000).

A common element of the arguments is that widespread glaciation occurred on land and that this led to

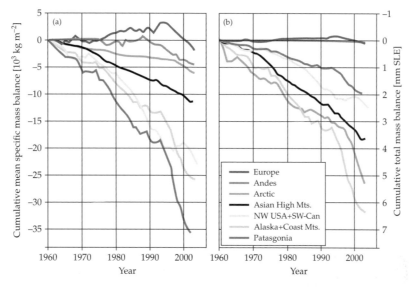

Figure 1.6 The decline in terrestrial ice mass in large regions (Dyurgerov and Meier 2005) as shown by (A) the cumulative mass balance (in terms of tonnes of ice lost per square metre) and (B) the cumulative effect on sea level (in terms of millimetres). The relative strength of the response of climate change by different regions is clearly shown in (A), with Patagonia losing most ice in terms of mass per unit area. By contrast, the glaciers of Alaska and the surrounding mountains have the largest effect on sea level rise (B). SLE, sea level equivalent. From *Climate Change 2007: the Physical Science Basis. Working Group I Contribution to the Fourth Assessment Report of the Intergovernmental Panel on Climate Change*, Figure 4.15, Cambridge University Press; used with the permission of the IPCC. See also Plate 2.

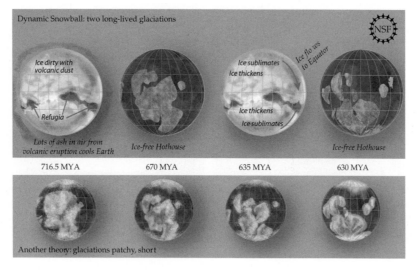

Figure 1.7 A schematic representation of the growth of ice on Snowball Earth, and the subsequent melting of the snowball as greenhouse gases (principally CO_2) build up in the atmosphere. Image by Zina Dertsky, National Science Foundation. See also Plate 3.

terrestrial eukaryotic phyla, including red, green, and chromophytic algae, being forced into and adapting to cold niches (Hoffman *et al.* 1998) such as cryoconite holes (Tranter *et al.* 2004). Recent work also suggests that microbial life will have survived beneath ice sheets if the energy balance of the sheets allowed water to exist at their beds (Hodson *et al.* 2008). Volcanoes, which either rose up through the ice sheets or erupted through them, injected CO_2 into the atmosphere that accumulated to high levels over time in the absence of runoff and terrestrial chemical weathering (Hoffman *et al.* 1998), which serves to remove CO_2 from the atmosphere (Holland 1978). The consequent increase in air temperature was eventually sufficient to melt the glacier ice and to reverse the runaway glaciation, since the terrestrial land surfaces were now able to adsorb solar radiation. Hence, the Earth flip-flopped between periods of very cold and very hot climates. Enhanced terrestrial chemical weathering during deglaciation would have added large fluxes of Ca^{2+} and HCO_3^- to the oceans. The consequent warming of the oceans caused wide-scale precipitation of cap carbonates, since carbonate is less soluble in warm water.

Research into the evolution of life on Earth thus has to consider how Snowball Earth impacted on the survival of life during these violent and persistent changes in climate during the Neoproterozoic. Indeed, even earlier life around 2 billion years ago might have had to evolve with more frequent climatic flip-flops (Runnegar 2000). Recent advances in our understanding of how the present-day cryosphere provides a range of refugia for microbial life make it less problematical that terrestrial life was able to survive under conditions of global glaciation, and that a wide range of terrestrial biodiversity was maintained in cryospheric habitats (Hodson *et al.* 2008).

1.4 Sea ice

1.4.1 Nature of sea ice

Extensive regions of the most southerly and northern seas and oceans are covered by sea ice. The extent of the cover changes seasonally, being more extensive in the winter (Figure 1.8). At its maximum, sea ice covers 13% of the Earth's surface and as such

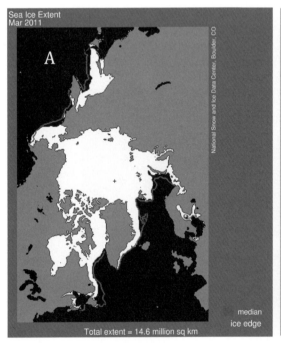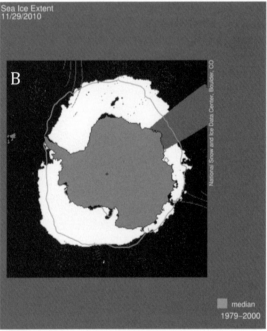

Figure 1.8 (A) Sea ice extent in the Arctic in March; (B) sea ice extent in Antarctica in November. Images courtesy of the National Snow and Ice Data Center, University of Colorado, USA.

represents one of the planet's major ecosystems. The largest extent of sea ice occurs in the Southern Ocean around the Antarctic continent (Thomas and Dieckmann 2002) (Figures 1.9 and 1.10).

Sea ice is not a solid mass but is made up of a matrix containing numerous channels and pores. These areas provide a habitat for complex communities of algae, protozoans, bacteria, and viruses, which in turn support metazoans like copepods and krill at the sea ice base and in the larger brine-filled channels. The extent of this network of channels has been investigated using a fluorogenic tracer technique (Krembs *et al.* 2000). This approach has elucidated the internal surface area of channels and the accessibility for organisms of different sizes under different temperatures and ice textures. The total internal surface area ranged from 0.6 to $4\,m^2\,kg^{-1}$ of ice and declined as temperatures decreased. At $-2°C$, 6–41% of the surface area of channels was covered by microorganisms. As cooling to $-6°C$ occurred, the coverage by microorganisms increased as they were concentrated onto a smaller surface area.

The morphology of the brine-filled web of channels has major implications for food web relationships in the sea ice. The colonization of the ice largely depends on the structure of the brine channel network. Over half of the brine channels are smaller than $200\,\mu m$ in diameter. Algae, protozoa, and bacteria inhabiting these narrow channels escape grazing pressure by larger predators. As in the soil, organisms with elongated, flexible body forms are best suited to traversing the narrow channels. For example, Krembs *et al.* (2000) found that rotifers possessed an extraordinary capacity to penetrate very narrow cylindrical channels and that ciliates had access to the majority of spaces (Figure 1.11). Approximately 50% of the entire internal surface of the brine channel network is less than $41\,\mu m$ in diameter. These very narrow channels are colonized by bacteria and smaller protozoans, such as heterotrophic and autotrophic nanoflagellates and small ciliates. However, despite these very narrow channels offering a refuge from predation by

Figure 1.9 Pack ice in the Southern Ocean. See also Plate 4.

Figure 1.10 The land-fast sea ice 30 nautical miles off Mawson Station in the Australian Antarctic Territory showing ship bows on the top right, Adelie penguins in the foreground, and mountains in the background.

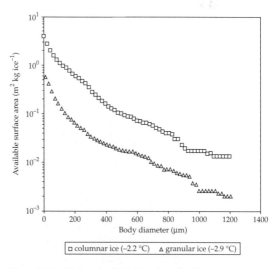

Figure 1.11 Maximum available brine channel surface area for predators of a given body diameter for sea ice at approximately $-2.5°C$. Data for granular ice and columnar ice are shown. Redrawn from Krembs *et al.* (2000).

larger ciliates and flagellates, they are unfavourable for the diffuse transport of molecules vital for sustaining physiological function.

Variations in the extent of the sea ice over many years have the potential to act as an important indicator of climate change. Since the 1970s we have had reliable satellite images that indicate that the extent of the annual sea ice around the Antarctic continent has remained stable, with only a slight shift in sea ice extent and the length of the sea ice period. However, 30 years is insufficient time to accurately assess long-term trends. There are other records on the extent of sea ice from whaling ship logs. The whaling fleet followed the retreat of the ice edge southwards each year in pursuit of their livelihood. Whales congregate at the proximity of the ice edge because the marginal sea ice zone has enhanced biological activity, particularly of krill. One of the major food sources for higher trophic levels in the Southern Ocean is krill, which has rightly been described as the mainstay of the Southern Ocean (Figure 1.12).

A mathematical analysis of whaling fleet data using detailed statistical methods reveals that there was a southward shift in sea ice extent between 1931/1932 to 1955/56 that was substantially greater than in the period between 1971/1972 to 1986/1987. The change was in the region of 1.89° to 2.80° of latitude with a mid-range estimate of 2.41°. There were regional variations, with the largest changes detected in the South Atlantic (de la Mare 2009).

There is other evidence of climate change on the continent and on the surrounding islands to support these data.

Evidence of changes in sea ice extent is also provided from investigations in the Arctic. A trend for warming has been evident during the last two decades, which has led to a reduction in the area and thickness of the Russian Arctic sea ice. This has created significant changes in the structure of sea ice communities and the under-ice water layer (Melnikov 2005). Off eastern Canada light ice years have been noted between 1996 and 2002 (Johnston *et al.* 2005). Heavy ice years correlate with positive North Atlantic Oscillation (NAO) conditions. The NAO is a large-scale fluctuation in atmospheric pressure between a subtropical Azores high and a polar Icelandic low and is a major factor in climatic variability in the North Atlantic. Years when the sea ice is 'light' have major implications for seal species that pup on the sea ice. The pups are born on the ice and have to be mature enough to survive in the sea before the ice melts.

1.4.2 Sea ice communities

Organisms that live within the sea ice face major physiological challenges. Temperatures are always low but can fall to –20°C within concentrated brine. Salinity variations are large, so that organisms must be capable of tolerating hypersaline conditions, but be equally tolerant of hyposaline conditions,

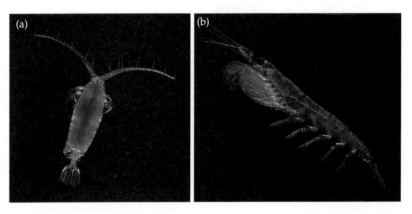

Figure 1.12   Marine crustaceans: (A) Aetideildae calanoid copepod; (B) Antarctic krill *Euphausia superba*. Photos courtesy of Russ Hopcroft, University of Alaska Fairbanks and CoML.

because when the ice starts to melt the brine is highly diluted by water with a low conductivity. When seawater freezes 70% of its dissolved salts are excluded. Ice is effectively close to the conductivity of freshwater. Thus the ability to withstand osmotic stress is an important adaptation.

Anyone who has travelled to the polar regions by ship in spring or early summer will have seen the underside of the sea ice as blocks of it are thrown aside by the bows of an ice-breaking ship. What is immediately striking is the layer of brown that covers the lower sea ice. This is a community dominated by diatoms (see Figure 1.4) that contain high levels of the brownish coloured photosynthetic pigment xanthophyll. Xanthophyll masks the green colour of the chlorophyll *a* that all algae possess. Chlorophyll *a* concentrations are widely used by aquatic scientists as an estimate of the biomass of photosynthetic communities of phytoplankton or sea ice communities. In the sea ice values for chlorophyll *a* can reach up to $1000 \mu g \, L^{-1}$, which in turn supports extremely high levels of photosynthesis or primary production. Ice attenuates light more sharply than water so that less than 1% of surface irradiance reaches the water immediately under the ice (Laurion *et al.* 1995). When the ice also has a snow cover, light becomes severely attenuated. Sea ice algae are among the most shade-adapted autotrophs on Earth. Their ability to carry out photosynthesis is inhibited at irradiances as low as $20 \mu mol$ photons $m^{-2} s^{-1}$ and photosynthesis has been detected at irradiance levels as low as $1 \mu mol$ photons $m^{-2} s^{-1}$ (McMinn *et al.* 2003). It is usually light levels not the availability of nutrients that limits photosynthesis in the sea ice. Thus in order to survive under such conditions the so-called ice algae have photosynthetic systems that are saturated at low levels of incident radiation and have evolved highly efficient photosynthetic systems. In the longer term they can adapt by changing their photosynthetic pigment composition, increasing the accessory pigment fucoxanthin, which is most efficient at capturing wavelengths that can penetrate the sea ice (Thomas and Dieckmann 2002).

Diatoms within the sea ice release ice-active substances called extracellular polymeric substances (EPS) made up of glycoproteins and polysaccharides. EPS act to change the environment surrounding the cells and probably buffers against salinity and pH changes as well as freezing effects. EPS are effectively dissolved organic carbon and contribute to the pool of DOC that supports the microbial loop that operates within the sea ice, as it does in the open waters of the sea and other aquatic environments. As shown in Figure 1.1, bacteria use the DOC pool as a source of energy. Various studies have shown that the highest densities of bacteria occur in the lower layers of the sea ice close to the ice–water interface. Bacterial abundances ranged up to 20×10^5 cells m^{-2}, while heterotrophic nanoflagellates (HNAN) that feed on the bacteria and phototrophic nanoflagellates (PNAN) (photosynthetic flagellates) ranged up to 25.9×10^8 cells m^{-2} (Laurion *et al.* 1995). The numbers depended on the degree of snow cover on the ice. One would normally expect higher numbers of bacteria than their predators, the HNAN. The authors attributed the discrepancy to the likelihood that the HNAN were also exploiting DOC as an energy source. There are numerous reports of HNAN being capable of taking up and using DOC to support growth. Interestingly, the net bacterial growth or production was only a small fraction of the co-occurring algal production. This is unusual, as in most aquatic environments bacterial production can be significant and may be up to 50% of primary or algal production. Low temperatures impact on the degree of bacterial production and this is likely to be the explanation in the sea ice.

Various studies in Antarctica have shown that certain PNAN species are particularly common in the sea ice, including *Pyramimonas* species, chrysophytes, cryptophytes, euglenoids, and *Mantoniella*. A number of these groups are capable of mixotrophy (see Figure 1.3).

HNAN, PNAN, and bacteria are subject to predation by ciliated protozoans. A study of sea ice in the Ross Sea showed that the diversity of the community was relatively small compared with the open waters, but included *Strombidium* species, *Mesodinium rubrum*, *Didinium*, and small bacterial feeding ciliates known collectively as scuticociliates (Stoecker *et al.* 1993). These are all species found commonly in the marine plankton. *M. rubrum* is a particularly interesting, highly adaptable ciliate (see Figure 1.2A). It sequesters the plastids of its cryptophyte prey and uses them for photosynthesis; thus it is an autotroph. It is ubiquitous in estuaries and seas from the equator to the poles and is one of the species that causes so-called red tides when it becomes especially abundant. Its presence in sea ice is not surprising; in the marine-derived saline lakes of the Vestfold Hills in Antarctica it is one of the most common ciliates and survives salinities as low as 4‰ up to hypersaline 60‰ (seawater is 35‰) (Perriss *et al.* 1995; Perriss and Laybourn-Parry 1997). The sea ice is a challenging environment where only the most robust of species can survive. *Mesodinium* was found in 90% of the samples collected by Stoecker *et al.* (1993).

Dinoflagellates are another conspicuous component of the sea ice protozoan community (see Figure 1.4C, D). The dinoflagellates are classified as phytoflagellates, but as indicated above around half of them do not contain photosynthetic pigments and are heterotrophic, feeding on DOC, bacteria, and other protozoans. Their nutrition is complex and some species are mixotrophic. Heterotrophic dinoflagellates constituted between 5% and 15% of the protozoan community of sea ice in the Ross Sea in summer, among which *Gymnodinium* predominated (Stoecker *et al.* 1993). One of the common autotrophic dinoflagellates found in the Antarctic sea ice is the small species *Polarella glacialis*. *Polarella* reached its highest density of $1.5 \times 10^6 \, \text{L}^{-1}$ of brine in mid-November in Prydz Bay off Davis Station (East Antarctica), declining into December when they encysted.

Little is known about viruses in the sea ice, but they do occur in abundances of 10^6 to $10^7 \, \text{ml}^{-1}$ of brine (see Figure 1.3D). A study of Antarctic sea ice showed that around 18% of the viruses observed were large and electron microscopy showed that they were infecting both HNAN and PNAN, in particular the autotrophic nanoflagellate *Pyramimonas tychotreta*. In the case of *P. tychotreta*, viral infection was implicated in causing the cessation of a bloom. Thus while some of the viruses were clearly infecting flagellates the majority were bacteriophage (Gowing 2003).

These populations of autotrophic and heterotrophic microorganisms provide a food source for a range of invertebrate metazoans that inhabit the sea ice, particularly the ice–water interface. Diatoms are particularly important to copepods and

euphasids, especially krill (Class Crustacea) (see Figure 1.12B). Krill are important grazers of primary production in the Southern Ocean and a critical prey source in the reproductive success of seals and seabirds. The life cycle of krill is complex. Embryos are released by females during the austral summer, following which the larvae develop through a series of juvenile stages (the early ones bearing no resemblance to the adult) within the following 11 months. Their first winter is a critical stage in survival. At this time food is scarce in the water column. During winter development is slowed. There is strong evidence to suggest that larval krill survival depends on exploiting food resources associated with the annual sea ice. Larval krill have been observed feeding on the ice undersurface in winter. Indeed, larval krill from the pack ice showed higher growth rates and higher lipid content when compared with larval krill collected from open water. A number of workers have shown that there is a strong correlation between the recruitment success of krill and the timing and extent of sea ice formation during winter (Siegal and Loeb 1995; Quetin and Ross 2003).

In a study that spanned winters it was shown that the growth rate was strongly correlated with day length. This is because krill larvae shift to the under-ice habitat in early winter where they live either in the sea ice or the water immediately below it. The correlation is probably related to light levels driving photosynthesis (krill food) both in the water column and the ice. Growth rates were close to zero in early winter, but by late winter had increased twofold. This growth pattern was consistent across winters (Quetin *et al.* 2003). The importance of krill to the Southern Ocean food web has made it the focus of much investigation. Clearly the sea ice and the food resources it offers are crucial in sustaining recruitment to the krill populations and in turn to the important fisheries in the Southern Ocean.

1.5 Lake ice

Many lakes in the polar regions, higher continental latitudes, and those located at altitude are covered by seasonal and in some cases permanent ice cover. The majority of lakes are freshwater, but saline marine-derived lakes do occur, particularly in Antarctica. Lake ice cover differs significantly

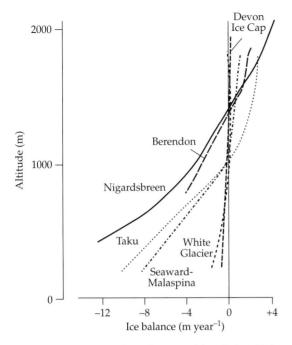

Figure 1.18 Altitudinal profiles in the net mass balance for low altitude glaciers in the Canadian High Arctic (Devon Island and White Glacier), Scandinavia (Nigardsbreen), and Alaska (Berendon, Seward-Malaspina, and Taku). Redrawn from Benn and Evans (2010).

ing is also disproportionately high nearer the margin since the influence of the ice surface upon air temperatures is likely to be lowest there. However, a key feature of the ablation of continental-scale ice sheets is the importance of calving, which tends to occur in locations of ice streams and major outlet glaciers. A recent assessment of mass loss from Antarctica emphasized the importance of calving loss from West Antarctica in particular whilst reporting a total calving flux of ~2193 Gt year^{-1} during 2000 (Rignot *et al.* 2008). This was in excess of the accumulation inputs, which were estimated to be 2055 Gt year^{-1}. The input and output mass fluxes estimated here seem trivial when distributed over the entire ice sheet surface, being 0.180 and 0.169 m w.e. year^{-1}, respectively.

1.6.2 Hydrological zonation in surface ecosystems

A major barrier to understanding the net mass balance of ice sheets is the fact that their size greatly

hinders the use of traditional glaciological monitoring methods. Much research has therefore resorted to the use of remote sensing techniques for ice sheet mass balance assessment. These allow the use of altimetry to detect changes in ice thickness, and spectroradiometry to assess the water content of snow and ice. The latter is of great importance to readers of this book, and results from the sensitivity of microwave emissivity to snow pack water content (Mätzler 1998). This method has promised satellite-derived estimates of snow pack water equivalent content for some time, but without satisfactory results (Rutter *et al.* 2009). However, it has delivered reliable estimates of the areal extent of different melt zones upon ice sheets (e.g. Fettweis *et al.* 2007), which can then be used to understand the locus and duration of enhanced surface biological activity. Figure 1.19 shows zones typical of the Greenland Ice Sheet (after Hall *et al.* 2009), which include dry snow, wet snow/slush, and bare glacier ice. The zones compare favourably with those deduced using ground-based observations upon smaller glaciers by glaciologists working in the 1960s (e.g. Muller 1961). Hanna *et al.* (2008) also showed how simple temperature index models driven by data sets from general circulation models can be used to simulate the growth of these zones upon the Greenland Ice Sheet. At this scale (Vaughan 2006), a climate-driven surface mass balance model needs to be coupled with a runoff retention submodel (Janssens and Huybrechts 1999) in order to establish when the snow pack can no longer store meltwater. Since this transition occurs close to the time when water and nutrients reach the base of the snow pack, this approach offers the benefit of modelling the onset of microbial activity at the snow–ice interface: something that remote sensing methods are unlikely to achieve if this occurs while deep snow cover persists.

Remote sensing methods have also been used to map the melt zone of the Antarctic. For example, Ridley (1993) has reported a significant increase in duration of melting on the Antarctic Peninsula over the interval 1979–1991 using passive microwave sensors. In addition, indirect estimations of increasing wet snow zones have been deduced from dry snow mapping using the Synthetic Aperture Radar (SAR) (Rau and Braun 2002). This SAR study showed that

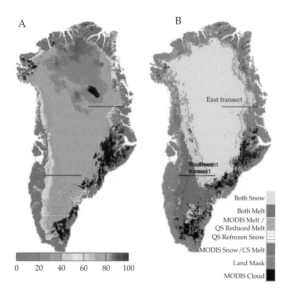

0   20   40   60   80   100

Figure 1.19 Extent of melting of surface snow and ice on the Greenland Ice Sheet during July 2007, as determined by satellite remote sensing. (A) The albedo of the surface is near 100% wherever there is cold snow, meaning that all incoming solar radiation is reflected back into the atmosphere. (B) As the snow starts to melt, the albedo drops, and when all the snow has melted to reveal the underlying grey ice, the albedo drops even further, down to as little as 30%. MODIS refers to data collected by Moderate-resolution Imaging Spectroradiometry, and QS refers to melt calculated from the Daily QuikSCAT Melt Special Product algorithm. More detail can be found in Hall et al (2009). From Hall *et al.* (2009), with permission of the American Geophysical Union. See also Plate 6.

north of 70°S, the summer melt extent for (non-ice shelf) terrestrial snow packs is typically ~30×10^3 km². The analysis was also able to detect a percolation zone, defined by the presence of refrozen ice lenses, which indicated that the maximum extent of melting is more likely to be 85×10^3 km². The extension of these analyses to continental Antarctica has revealed a limited extent of surface melting south of 70° latitude (Torinesi *et al.* 2003). However, these methods are likely to underestimate the extent of biological activity due to the importance of subsurface melting. The contribution of subsurface melting to total areal melt in Antarctica was therefore assessed using a meteorological model by Liston and Winther (2005), who found that 11% of the total of 422 km³ year⁻¹ of meltwater production took place beneath the surface (Table 1.2).

Subsurface melting is well known upon ice surfaces kept snow-free and polished by katabatic winds. These are so-called 'blue ice areas' (BIAs), which cover between 1% and 2% of the Antarctic Ice Sheet surface (Bintanja 1999; Winther *et al.* 2001). Four types of BIA have been established by Takahashi *et al.* (1992) according to the manner by which the windfield interacts with the ice surface to remove the snow (Figure 1.20). For example, types 1 and 2 are formed by airflow over or around mountains that protrude through the ice (nunataks), whilst

Table 1.2 Antarctic area experiencing melt, and the associated blue ice area fraction and total Antarctic area fraction (based on a blue ice area of 239 902 km² and an Antarctic area of 13 746 463 km²).

	Area (10⁶ km²)	Blue ice fraction (%)	Antarctic fraction (%)	Meltwater production (km³ year⁻¹)
Blue ice				
Surface melt	0.10	42.4	0.7	2.0
Subsurface melt	0.19	79.1	1.4	57.4
Snow				
Surface melt	1.53	–	11.1	46.0
Subsurface melt	2.77	–	20.2	316.5
Total				
Surface melt	1.63	-	11.8	48.0
Subsurface melt	2.96	-	21.6	373.9

Also shown is the ten-year-average annual meltwater production.
After Liston and Winther (2005).

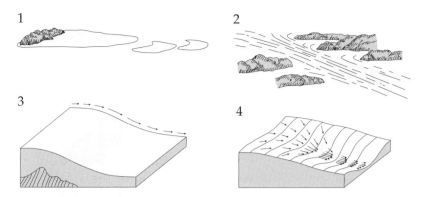

Figure 1.20 Four types of blue ice areas: types 1 and 2 are formed by mountains protruding through the ice, causing eddy-like circulations (type 1) and locally concentrated air flows (type 2). Types 3 and 4 are caused by ice surface slope changes, causing airflow acceleration (type 3) and locally concentrated air flows in valleys (type 4). Redrawn from Takahashi *et al.* (1992).

types 3 and 4 are the consequence of wind acceleration by ice surface topography alone. These differences are important from an ecological perspective, since the former types often supply debris to which microorganisms can become attached and around which the probability of melting is greatly increased due to a reduction in surface reflectance (albedo) (see Rasmus, 2009). Chapter 3 will show how the translucence of blue ice is particularly important for allowing photosynthesis by microorganisms associated with such debris, referred to as 'cryoconite' (see below). Thus primary production in a watery habitat is able to occur in spite of an overlying ice 'lid' (Tranter *et al.* 2004), so habitats not unlike those found in the perennial ice covers of polar lakes (Section 4.2) are typical of many blue ice areas.

1.6.3 Supraglacial lakes

Supraglacial pools and lakes that form upon glaciers during the summer often become ephemeral, oligotrophic ecosystems (Vincent 1988; Hawes *et al.* 2008). Of particular importance on melting glaciers are the small, debris-filled pools called 'cryoconite holes' (Figure 1.21), whose biological activity has been studied across a worldwide selection of glaciers. Their formation is a consequence of the low albedo of the debris, allowing solar heating to melt the debris into the ice and resulting in a small, cylindrical melt pool. Cryoconite debris is also found in larger lakes upon polar glaciers (Fountain *et al.* 2004) and ice sheets (e.g. Box and Ski 2007). The

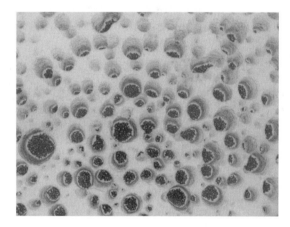

Figure 1.21 A myriad of small cryoconite holes upon the Vestfonna Ice Cap in Nordaustland, Svalbard.

gentle ice surface slopes promote the formation of extensive yet shallow lakes when conditions are favourable for melting (Sneed and Hamilton 2007; McMillan *et al.* 2008). For example, a study of West Greenland revealed lake areas up to 9 km², yet the maximum depth was only ~12 m (Box and Ski 2007) (Figure 1.22). Since cryoconite debris has been detected in these lakes, they offer a basis for remotely sensing those parts of the ice sheet surface where smaller cryoconite holes are also present.

Supraglacial lakes are often ephemeral because they drain vertically through crevasses or holes called 'moulins' after some threshold water pressure has been exceeded (e.g. Hambrey 1984) (Figure 1.23). At the time of writing, researchers are trying to link these events to

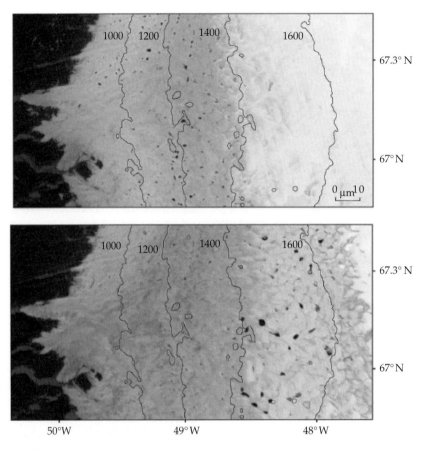

Figure 1.22 Supraglacial lakes near the southwest margin of the Greenland Ice Sheet. Lakes range in size from a few tens or metres to a kilometre or so across. The upper satellite image was taken on 11 June, whereas the lower satellite image was taken on 16 July. Note how, in the latter, there are more supraglacial lakes within the region from which the overlying snow pack has melted. The black lines denote the altitude of the ice surface. From Sundall *et al.* (2009), reproduced with the permission of Elsevier. See also Plate 7.

speed-up events upon the Greenland Ice Sheet (e.g. Das *et al.* 2008), because their sudden drainage over-pressurizes the subglacial drainage system. In so doing, a basis for 'injecting' surface-derived nutrients and microorganisms into subglacial sediments takes place in a manner similar to extreme rain-on-ice events (Hodson *et al.* 2009). Moderate-sized Himalayan supraglacial lakes also display sudden drainage characteristics and in fact represent a significant hazard on account of their high elevation position in steep, sometimes populated, catchments (Reynolds 2000). These lakes tend to form upon glaciers with substantive debris cover (average thicknesses between 10 and 70 cm in Konovalov 2000) and represent a largely unexplored supraglacial ecosystem at the time of writing.

1.6.4 Water distribution in subsurface ecosystems

Seasonal and diurnal changes in climate greatly influence microbial activity in ice surface ecosystems due to the onset of freezing. However, beyond ~15 m depth, these variations are muted (Paterson 1994) and so spatial variations in ice temperature and water content become more important than temporal variations, unless the entire glacier profile lies at the melting point (see below). Here there are also several other abiotic factors that greatly limit biological productivity. Firstly, there is insufficient light penetration to enable photosynthesis. Secondly, ice overburden pressure increases with depth and many

Figure 1.23 A moulin capturing water from a supraglacial stream on the Haut Glacier d'Arolla, Switzerland, and transmitting it to the glacier bed. The backpack in the top left is included as a scale reference.

glaciers are in excess of 100m thick, whilst much of the East Antarctic Ice Sheet is in excess of 2000m thick. Thirdly, the rejection of solute (and microbial cells) takes place during interstitial freezing (Price 2000; Mader *et al.* 2006), which tends to result in acidic conditions (Wolff and Paren 1984), or at the very least, highly concentrated solutions. Finally, glaciers are effectively 'self-sealing' (Lliboutry 1996), so that the interstitial percolation of surface-derived nutrients, microorganisms, and water into the glacier is almost impossible. Therefore habitats entombed at depths in excess of 15m seem far more unlikely to support biological production than the glacier surface during the summer. Further, the resident microorganisms are almost certainly likely to have arrived there by slow accumulation processes rather than rapid hydrological transport. This prolonged storage maximizes the potential for cell damage. However, several works show that these cells can maintain their viability, even for millennia (e.g. Abyzov 1993). To understand this, Price (2000) has examined the crystal-scale hydrology of ice, giving particular attention to the relationship between solute concentration, ice crystal size, and the diameter of their interstitial veins. This shows that motile microorganisms are easily capable of metabolism in vein networks a few microns in diameter when the ice temperatures are no more than a few degrees below the melting point. Similar-sized water layers at the surface of clay particles within ice have also been proposed as another 'englacial habitat', this time enabling bacterial metab-

olism at even lower temperatures than interstitial veins (Price 2007). More recent work has even begun to question the need for motility within water, suggesting that diffusion through ice can in fact supply sufficient nutrients to enable metabolic activity for entombed bacteria within the ice crystal itself (Price 2009). These theoretical habitats seem plausible and ice core analyses are beginning to yield important evidence in their support (Priscu *et al.* 2006). *In situ* measurements of biological production remain elusive though, and it seems most likely that they simply enable 'survival metabolism' (Price 2009). After all, the physical conditions are harsh and there is a lack of nutrient supply due to the lack of hydrological connectivity to the surface environment (Lliboutry 1996).

The melting point of glacier ice is depressed by ice loading at a rate of by about 0.07°C per 100m (Paterson 1994). However, since not all glaciers lie at the melting point, glaciologists have used the concept of *thermal regimes* to understand the complexity that results. Several key thermal regimes may be defined: (i) glaciers whose subsurface temperature is entirely beneath the pressure melting point (hereafter a *cold ice* thermal regime); (ii) those entirely above it (at least for ice below the surface layer, and hereafter a *warm ice* thermal regime); and (iii) those where pressure melting is achieved at greater depths (hereafter a *polythermal ice* regime). Figure 1.24 shows the types of temperature variations that can be expected at depth for cold and polar ice masses. The presence of pressure melting in warm ice renders the interstitial and particle surface habitats described by Price (2009) more favourable for biological production, but the impermeability of such ice remains (Lliboutry 1996) so nutrient supply is still a major issue. In fact, Lliboutry has shown that only when ice crystals are small and bubble-free will the permeability of warm ice become significant. Such conditions are more likely at the glacier bed, although other processes tend to dominate the drainage of water here as the following discussion will reveal.

1.6.5 Water in subglacial habitats

The presence of warm ice at the bed of glaciers greatly facilitates glacier flow, leading to a key process that allows the ingress of water and nutrients: crevassing. Cold-based glaciers show by far the lowest incidence

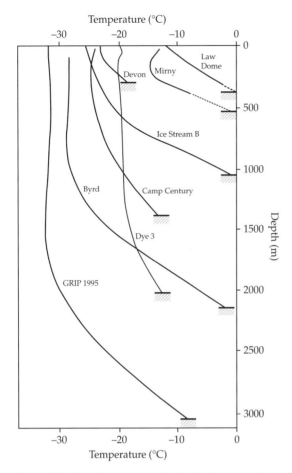

Figure 1.24 Vertical temperature profiles though polar ice sheets. Note that the pressure melting point of ice decreases roughly by 0.7°C for every kilometer of ice cover. However, most of the subsurface temperatures remain below the melting point. Redrawn from Benn and Evans (2010) and Clow et al. (1996).

of crevassing on account of their very low annual ice velocities. These are mostly achieved by the deformation of the ice under its own weight and are typically of the order of ~1m year^{-1} for polar valley glaciers. By contrast, warm-based valley glaciers can show velocities more than two orders of magnitude greater due to the additional processes of basal sliding and the deformation of subglacial sediments (Benn and Evans 2010). This results in a far greater incidence of crevassing. Interestingly, the annual velocities of polythermal-based glaciers are intermediate to these extremes, usually because cold ice is found at the thinner margins of the glacier (Sugden and John 1984). Therefore, almost none of the water, nutrients, and innoculi liber-

ated by melting at the surface of cold-based glaciers are able to access the wet sediments distributed across the glacier bed. The corresponding proportion for warm-based glaciers can be as high as 100% (Hodson et al. 2008) and arterial subglacial networks can therefore develop beneath these glaciers during the summer (Fountain and Walder 1998). However, these require sustained meltwater inputs to offset destruction by ice deformation (Richards et al. 1996). As a result, subglacial drainage systems tend to be least well developed at the start of the summer, a time when the influence of microbial activity can be most obvious in the outflow due to prolonged residence times at the bed (Wynn et al. 2006).

The distribution of subglacial water beneath glaciers and ice sheets is poorly understood, although there has been some success predicting the location of the arterial drainage channels beneath temperate glaciers composed entirely of warm ice (e.g. Sharp et al. 1993). Tranter et al. (2005) and Hodson et al. (2008) have suggested that these channels provide important nutrients for high numbers of bacteria within subglacial sediments along their flanks (Sharp et al. 1999). However, finding such channels beneath the great ice sheets is proving difficult. First, great volumes of meltwater are required to maintain them beneath more than 400m of glacier ice (owing to high rates of deformation) (Benn and Evans 2010). Second, existing geophysical evidence for discrete channels are equivocal (Das et al. 2008). However, our understanding of water distribution beneath ice sheets is being greatly improved by the investigation of Antarctic subglacial lakes, which were first discovered by Robin et al. (1970). The recent employment of modern remote sensing methods has greatly advanced our understanding of the volume, geography, and hydrological dynamics of these habitats. Presently, at least 30000 km^2 of the ice sheet is likely to be underlain by subglacial lakes (Siegert et al. 2005). For comparison, the surface melt zone of Antarctica is thought to be ~1 630 000 km^2 (see Table 1.2), whilst the zone of subglacial warm ice is probably even greater (Figure 1.25). Recent research has demonstrated that these lakes are supplied by basal melting alone, yet movement of this water through a dynamic, channelized drainage system can even be detected upon the ice surface, some 3000m above (Wingham et al. 2006). Detailed biogeochemical and microbiological

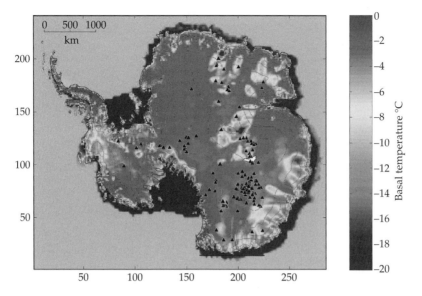

Figure 1.25 Basal temperature of the Antarctic Ice Sheet derived from the thermo-mechanical ice sheet model. Redrawn from Siegert *et al.* (2006). Triangles represent known subglacial lakes. See also Plate 8.

studies in one isolated Antarctic subglacial lake and one Icelandic subglacial lake (associated with a volcanic caldera) are notable and will therefore receive significant attention in this book (Priscu *et al.* 1999; Gaidos *et al.* 2004). These lakes (Vostok and Grimsvötn) provide virtually all that is known of contemporary subglacial lake habitats, although Hodgson *et al.* (2010) have shown that former subglacial lakes in the forefields of retreating glaciers can also provide valuable study sites, especially when they maintain a lake ice cover that prevents equilibration with the atmosphere. Interestingly, subglacial lakes beneath the Greenland Ice Sheet have remained conspicuous by their absence while writing this book.

Geophysical methods that were responsible for the discovery of Antarctica's subglacial lakes offer the most promise for improving further our understanding of how water is distributed beneath Earth's ice sheets. These methods include ground penetrating radar (GPR), which is ideal on account of the strong dielectric contrasts between cold ice, warm ice, and either partially or completely water-filled channels (e.g. Pettersson *et al.* 2004). Recent use of this technique has therefore improved our understanding of where basal melting occurs beneath the Greenland Ice Sheet (Oswald and Gogieni 2008). The use of active seismic methods in conjunction with GPR offers even more promise, since greater understanding of the distribution of wet subglacial sediments can be derived (Smith 1997) and such sediments are known to contain the most microbial biomass beneath both glaciers (Sharp *et al.* 1999) and Antarctic ice streams (Lanoil *et al.* 2009). However, these techniques have not been employed extensively and there are no reliable estimates yet of the expanse of these ecologically significant, wet sediments beneath Antarctica. Therefore our understanding of the distribution of water and water-rich sedimentary habitats is best developed in the case of certain small glaciers (e.g. King *et al.* 2008). The alternative approach, wherein models are used to assess the incidence of basal melting beneath ice sheets, is not trivial but holds promise with respect to simple two-dimensional (centre line) modelling (Wilch and Hughes 2000) and perhaps more sophisticated thermomechanically coupled three-dimensional models such as that depicted in Figure 1.25.

1.6.6 Overview: broad structure and characteristics of glacial ecosystems

The preceding review and earlier published syntheses (Fogg 1988; Hawes *et al.* 2008; Hodson *et al.* 2008)

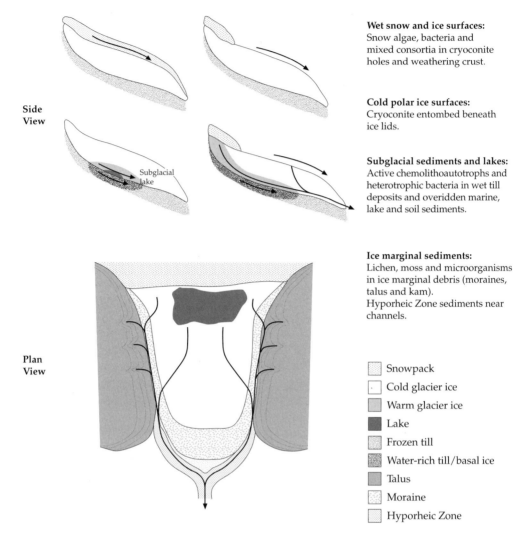

Wet snow and ice surfaces:
Snow algae, bacteria and mixed consortia in cryoconite holes and weathering crust.

Cold polar ice surfaces:
Cryoconite entombed beneath ice lids.

Subglacial sediments and lakes:
Active chemolithoautotrophs and heterotrophic bacteria in wet till deposits and overidden marine, lake and soil sediments.

Ice marginal sediments:
Lichen, moss and microorganisms in ice marginal debris (moraines, talus and kam).
Hyporheic Zone sediments near channels.

Side View

Subglacial lake

Plan View

Snowpack
Cold glacier ice
Warm glacier ice
Lake
Frozen till
Water-rich till/basal ice
Talus
Moraine
Hyporheic Zone

Figure 1.26 Different habitats within glacial ecosystems.

present a case for supraglacial, englacial, and subglacial habitats within the glacial biome (Figure 1.26). Most biological activity can be anticipated where water-rich sediments provide a stable substrate, and subsequent sections of this book will emphasize the attachment process that allows microorganisms to make use of this advantage. The outcome is that melting glacier and snow surfaces, which receive exotic debris from the atmosphere (dust) and energy from the sun, are some of the most active and expansive ecosystems in the cryop-

shere. However, strong regulations upon their activity are imposed by the duration and extent of melting, which the preceding text has outlined in some detail. A further expansive habitat is the glacier bed, where pressure melting and subglacial sediments support a dark, largely heterotrophic ecosystem. Between the surface and the bed, the englacial vein network represents perhaps the most extreme habitat, where pH, pressure, and high ionic strength in vein networks greatly compromise biological activity in hydrologically isolated micro-

Table 1.3 Habitats for life in glacial environments.

Habitat	Physico-chemical environment	Description
Melting snow surfaces	Ionic strength $\sim 10^{-5}$ M, pH 5–6, atmospheric pressure, high light levels	Melting snow surfaces enable photosynthesis and heterotrophic microbial activity. Globally, this is the most expansive habitat upon glaciers and ice sheets, at least at the start of the summer period
Subsurface slush and superimposed ice	Ionic strength $\sim 10^{-4}$ M, pH 5.5–7, atmospheric pressure, intermediate to dark light levels (snow cover dependent)	Biological activity is initiated when water, nutrients, and cells are leached to the base of the snow. The leaching provides a nutrient-rich environment that is most likely dominated by heterotrophic activity due to darkness. Proximity to ice surface debris increases productivity, pH, and biomass
Melting bare glacier ice	Ionic strength on glaciers 10^{-5} M, pH 6–7, high to intermediate light levels, atmospheric pressure until freezing initiated	The bare ice surface of glaciers is characterized by organic-rich debris known as cryoconite and inorganic-rich moraine deposits. Both of these are revealed as heterogeneous deposits following snow retreat across glaciers. Cryoconite supports significant autotrophic production, as well as heterotrophy. Moraine sediments upon glacier surfaces seem to favour heterotrophy because they are thicker. Debris-covered glaciers therefore most likely promote net heterotrophy
Englacial ice	Pressure >235 kPa, pH 0–5 (assuming acid aerosol present), no light, ionic strength up to 1 M	Interstitial veins can become extreme environments. Here 'survival metabolism' by heterotrophs is most likely. Less extreme conditions exist in so-called warm ice
Subglacial ice/sediment	Pressure >1000–30 000 kPa, pH 7–8, ionic strength 10^{-4} to 10^{-2} M, no light. Anoxia, at least in patches	Saturated subglacial sediments and sediment-rich basal ice promote significant heterotrophic activity (and perhaps chemoautotrophic primary production)
Subglacial lakes	As above, but there is major uncertainty over REDOX conditions, which could range from superoxygenated to fully anoxic. Geothermal influences may also exist, with ionic strength and composition then dependent upon meltwater dilution	Lake water, sediments, and accretion ice, formed by the refreezing of lake water, offer potential habitats for heterotrophic and chemolithoautotrophic production. Some lake sediments and microorganisms might be formerly marine

niches. It is therefore fascinating to consider how subglacial microorganisms beneath the great polythermal ice sheets, especially in subglacial lakes, are able to become biologically active following transit through the englacial environment.

Table 1.3 shows the range in abiotic conditions that can be expected in glacial habitats. This reveals how the most extreme abiotic conditions for life can be anticipated in the vein networks of cold ice. Here acidic interstitial water is very likely (pH of approximately 1), whilst in saturated subglacial tills pH values of approximately 8 can be expected following rock–water interaction. Thus a variation in H^+ concentration of eight orders of magnitude can be expected in the case of polythermal ice sheets with subglacial melting. Ionic strength is most dilute

upon the ice surface when melting takes place because glacier ice is formed in many cases by snow that has lost much of its solute content by elution. Thus slush and superimposed ice formed from it can be nutrient-rich relative to ice melt.

In terms of the expanses of glacial habitats, melting snow is almost certainly the most important. However, many more data are required before an accurate assessment of the distribution of biologically active, saturated till beneath the polar ice sheets is known with any confidence. Estimates of maximum melting snow surface areas on the Greenland and Antarctic sheets are $\sim 0.86 \times 10^6$ and 1.53×10^6 km^2, respectively, using the sources reviewed above, but the figure is impossible to estimate for the multitude of glaciers elsewhere in the cryosphere. In these cases, however, the maximum total ablation area can be estimated (~0.48×10^6 km^2 according to Anesio *et al.* 2009). This provides a minimum estimate of the extent of melting upon glaciers, although it is most likely to be representative of the maximum area of bare glacier ice that is exposed at the end of summer. When melting, this habitat is most likely dominated by cryoconite ecosystems. Upon the ice sheets, bare glacier ice represents about 0.25×10^6 km^2 by the end of the summer in Greenland (Hanna *et al.* 2008) and 0.10×10^6 km^2 in Antarctica using the surface melt estimates from Liston and Winther (2005). This and other work shows that the largest single melting surfaces in Antarctica are its ice shelves, representing ~1.2×10^6 km^2 of snow and ice that are well known for surface lake development (Hawes *et al.* 2008). Since sea ice is incorporated into these ice shelves, the geochemical conditions here can be quite different to those observed upon glaciers (Wait *et al.* 2009). Ice shelves of the northern hemisphere are almost entirely composed of sea ice and represent similar habitats of up to 1043 km^2 (Mueller *et al.* 2006).

1.6.7 Life on glaciers

It is evident from the previous section that glaciers offer a range of habitats for colonization. The surface, and in particular cryoconite holes (see Figure 1.21), have been a focus of attention for many decades. Most of the early studies were qualitative, providing a description of what lives in the water contained in the holes and in the 'cryoconite' sediment (Steinbock 1936; Gerdel and Drouet 1960; Wharton *et al.* 1981). Cyanobacteria are a major constituent, as well as flagellates, many belonging to what are commonly know as the 'snow algae', other protozoa, diatoms, rotifers, and fungi. In the more extreme Antarctic glaciers, the cryoconite hole communities were entirely dominated by Bacteria and Protozoa (Wharton *et al.* 1981).

One of the first studies to quantify the communities and to measure processes was that of Säwström *et al.* (2002) on a High Arctic glacier in Svalbard. That study revealed bacterial concentrations between 1.0×10^4 and 3.6×10^4 bacteria mL^{-1} in the water and between 2.65×10^4 and 11.77×10^4 bacteria mL^{-1} in the sediment. Viruses also occurred in the cryoconite holes at concentrations of up to 12.71×10^4 mL^{-1} in the overlying water. Both heterotrophic and autotrophic nanoflagellates occurred (up to 400 cells m L^{-1}) as well as ciliates and rotifers. Photosynthesis in the holes was relatively high at up to 3.76 mg C L^{-1} day^{-1}, which is higher than that recorded in Antarctic and High Arctic lake plankton. Bacterial growth or production, however, was 86% lower than in neighbouring lakes, reaching 51 ng C L^{-1} h^{-1} (Anesio *et al.* 2007). Rates of bacterial production in cryoconite holes on glaciers in the Antarctic Dry Valleys were lower at 1.5 ng C L^{-1} h^{-1} in the overlying water. In the sediment rates were significantly higher at 138 ngL^{-1}h^{-1} (Foreman *et al.* 2007). Antarctic cryoconite holes are usually entombed, that is covered by ice and not in contact with the atmosphere, whereas in the Arctic and alpine locations the holes are open to the atmosphere and during periods of high melt may be in hydrological contact with each other (see Figure 1.21).

In recent years interest in cryoconite holes has grown and there is now a growing wealth of data, some of which elucidates the molecular biodiversity of the bacterial communities. In Dry Valleys cryoconite holes a high percentage of Cytophage-Flavobacteria were evident in the sediment, while in the overlying ice Betaproteobacteria cells predominated. A culture-based study also found Betaproteobacteria (Christner *et al.* 2003).

As indicated above, the base of glaciers offers another environment for biological activity. These environments are much more challenging to sample compared with the glacier surface. The subglacial environment is dark and lacking or low in oxygen. However, it contains high levels of carbon as glaciers and ice sheets override soils and peat deposits. Pioneering work in this environment indicated that it supported a community of largely psychrophilic (cold adapted) bacteria including aerobic chemoheterotrophs, anaerobic nitrate reducers, sulphate reducers, and methanogens (methane producing) (Skidmore *et al.* 2000).

Supraglacial lakes are frequently observed on glaciers. We have very little information on their communities but they are likely to be similar to cryoconites and Antarctic lakes and to be typified by simple truncated food webs, dominated by microorganisms. Since they are often ephemeral there is limited scope for colonization. Subglacial lakes, of which there are many under the Antarctic Ice Sheet, are as yet unexplored and offer a tantalizing picture as they have been separated from the atmosphere for millions of years. The best know is Lake Vostok, from which we have accretion ice from a drill hole above the lake. Accretion ice is found immediately above the lake and is derived from its waters, consequently it may be expected to contain organisms from the lake. Molecular analysis of the accretion ice revealed bacteria that were closely related to extant members of the Alpha-, Beta- and Gammaproteobacteria, Firmicutes, Bacteroidetes, and Actinobacteria. This suggests that the bacterial community of Lake Vostok may not be an evolutionary distinct biota (Priscu *et al.* 2008).

1.7 Snow

1.7.1 Physical and chemical characteristics

1.7.1.1 Formation of snow
Snow forms in the atmosphere when ice nuclei initiate the freezing of supercooled droplets (tempera-

tures can go down to –40°C) in clouds (Chernoff and Bertram 2010). Ice nuclei, which range in size from submicrons to a few tens of microns, are relatively few in number, and have compositions that range from inorganic sea salt, clays, and other minerals, through organic compounds formed in the atmosphere, to biogenic material including microbes with surfaces that promote ice nucleation (Fornea *et al.* 2009; Petters *et al.* 2009; Wiacek and Peters 2009). The snow crystal shape that grows for kinetic reasons is dendritic or rod-like, and is not thermodynamically stable (Chen and Baker 2010, Figure 1.27). The dendritic flakes are excellent scavengers of atmospheric particulates during their fall, and a host of different aerosols, including biogenic material such as fungal spores, pollen, and soil debris, are scavenged depending on the prior trajectory of the air mass through which the snow falls (Davies *et al.* 1984).

Snow flakes may melt, either partially or completely, during their fall, depending on the ambient air temperature. Then the snow flakes become rounded. They may refreeze if they are re-lofted by strong updrafts, for example in the air cells surround-

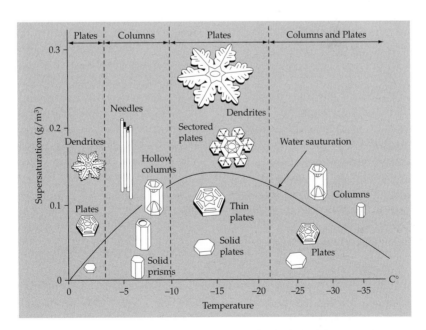

Figure 1.27 Main types of snow crystals, and their dependence on the temperature within the cloud and the amount of water present for their growth. From SnowCrystals.Com.

ing cumulonimbus clouds, and grow by accretion of further supercooled water droplets (Takahashi *et al.* 1999). Successive cycles of falling and re-lofting gives rise to hailstones, which include ice pellets known as groupel, which are solute-rich (see below).

1.7.1.2 Global distribution of permanent and seasonal snow cover

Cold, dry snow pack blankets most of Antarctica for much of the year because temperatures seldom reach 0°C even during the height of the summer. Snow cover eventually metamorphoses to glacier ice as a consequence. The same is true in the higher elevation accumulation areas of the Greenland Ice Sheet (see Figure 1.19) and the ice caps, although some summer melting and subsequent refreezing at depth can occur and is a common feature of the snow cover profile at lower elevations in the accumulation zone. The latter is also common in the snow cover on smaller glaciers worldwide, and climate change has resulted in the almost complete ablation of annual snow cover on glaciers in the Alps. Elsewhere, seasonal snow cover blankets the ground of large tracts of the northern hemisphere above 50°N for days through months each year. The typical depth is 20–180 cm in polar regions and continental interiors. Seasonal snow cover also occurs at higher elevations in South America, South Africa, Australia, and New Zealand (Gray and Male 1981).

1.7.1.3 Snow metamorphism in cold, dry snow pack

Dendritic snow flakes falling through cold air reach the ground and start to metamorphose almost immediately. Broadly, there are two types of metamorphism in dry, cold snow packs (Colbeck 1991). These are temperature gradient metamorphism (temperature gradients in the snow pack are greater than ~10°C/m) and equitemperature metamorphism (gradients are less than ~10°C/m). Equitemperature metamorphism produces a thermodynamically stable configuration of rounded grains bonded together where they touch by rounded 'necks'. Water vapour pressures are relatively high surrounding the highly convex sharp tips to the dendrites and are relatively low along the flat edges. Hence, water vapour leaves the tips (sublimation) and is deposited along the flat edges. Over

time, the grains round as the convex tips disappear and the flat edges grow. The areas where the grains touch form concavities, and water vapour pressures are lowest in these. Hence, some water vapour is deposited in the concavities, so forming necks that bond the grains together and increase the strength of the snow layer within which the process is occurring (Blackford 2007).

The ground over which snow lies in many parts of the northern hemisphere above 60°N, although often frozen when snow blankets it, is seldom as cold as the ambient temperature at the depth of winter. For example, many small mammals survive and breed beneath the snow cover that blankets the boreal forest in Canada over winter at temperatures of near zero at ground level, despite the ambient air temperature being over 20°C colder, as long as the snow depth is >30 cm so that the winter cold wave does not reach ground level (Gray and Male 1981). The fact that the ground is usually insulated from the overlying cold air by the snow pack means that there is usually good potential for temperature gradient metamorphism (TGM) to occur at the base of the snow pack in particular. TGM involves the transfer of water vapour from a relatively warm surface, for example from soil or the basal snow layers, to a relatively cold surface, usually cold snow layers at some distance above the base of the snow pack. The snow crystals that grow during TGM have shapes that are kinetically controlled, and are usually large and flat sided. These so-called faceted crystals are very poorly bonded and hence the associated snow layer is weak and avalanche prone. The faceted crystals formed are warmer than the crystals above, since latent heat is released during crystallization, so these pass on water vapour to the layers above and then cool. They are rewarmed by crystallization of water vapour from below, and pass on water vapour to the layers above. These successive transfers of water vapour are referred to as 'hand to hand' in the older literature. The net result is that many snow packs overlying alpine and Arctic soils contain weak basal layers of faceted crystals, often referred to as depth hoar. These basal layers may also be found on alpine and Arctic glaciers, although they are often flooded with snow melt during the melt season, which may refreeze to form superimposed ice, dependent on the temperature of the glacier surface. Surface hoar may also form on the surface of cold packs when warm, moist air overlies cold snow pack (Colbeck 1991).

1.7.1.4 Snow pack hydrology and snow metamorphism in wet or 'ripe' snow pack

Spring heralds the melting of seasonal snow pack. Snow melt at the surface percolates into the snow pack, either slowly as diffuse bands or quickly in preferential flow paths known as flow fingers. The first snow melt often freezes at depth in cold snow, so increasing the temperature of the lower snow due to the release of latent heat (Marsh and Woo 1984, 1985). It requires about the same amount of energy to melt a unit volume of snow as is required to heat that same volume from 0 to 100°. The ice layers that form at depth preclude the passage of diffuse melt flow, and so promote the formation of flow fingers in areas where the ice layers first melt out or pinch out laterally (March and Woo 1984, 1985). Percolation of snow melt through the snow pack transfers energy to the lower colder layers, and warms the pack until it eventually reaches 0°C throughout, when it is called equitemperate, warm, or ripe snow pack. Water reaching the base of snow pack on glaciers often refreezes and forms superimposed ice. Water reaching the base of snow pack on frozen ground refreezes to form an impermeable basal layer to the pack, which means that further snow melt is forced to flow laterally above the snow pack base, out of contact with the soil below.

Snow crystals metamorphose quite quickly when liquid water is present in the snow pack. Smaller grains melt and larger grains grow as a consequence. Hence, the snow pack becomes coarser grained during melting (Colbeck 1991). This in turn promotes the easier passage of water and the promotion of flow fingers. For example, snow crystals are usually larger around vegetation in the snow pack of boreal forests because of TGM and the effects of differential adsorption of radiation. This in turn results in flow focusing during snow melt, which may also enhance the transfer of nutrients from the snow cover to the vegetation below (Gray and Male 1981).

1.7.1.5 Chemical composition of snowfall

This depends on factors such as the origin of the air masses through which the snow is falling, the altitude at which the snow is deposited, and the meteorological conditions during snowfall (Davies *et al.* 1992). For example, maritime air masses will give rise to snow containing high concentrations of sea salt, while polluted air masses from industrial areas will deposit snow that is highly acidic due to the presence of strong acid anions (NO_3^- and SO_4^{2-}) from fossil fuel combustion (Suzuki 1987). Table 1.4 shows the chemical concentrations of snowfalls in different regions of the world. The chemical composition of snow shows spatial variability over distances of metres through tens of kilometres, reflecting factors such as the proximity to pollution sources, the impact of wind redistribution, and

Table 1.4 Chemical composition of the snowfall at selected sites throughout the world (after Davies et al. 1991).

Location and reference	pH	Ca	Mg	Na	K	NH$_4$	NO$_3$	SO$_4$	Cl
European Alps[1]	4.4–5.3	18–49	3–15	3–27	1–6	17–60	12–46	28–86	8–32
Central Asian mountains[2]	*	19–70	*	1–44	*	*	2.9–60	2.2–51	1–32
Turkey Lakes watershed, SE Canada[3]	4.57	34	0.9	10	0.2	7.5	19	17	3.7
Mid-Wales[4]	3.9–4.5	4–14	4–11	13–30	1–5	*	11–64	16–78	21–69
Sapporo, Japan[5]	4.4–6.4	13–63	18–67	59–190	2.3–6.4	*	*	70–99	63–310
Cairngorms, Scotland[6]	4.4	2.5	11	52	2.1	9.8	20	26	91
Svalbard[7]	5.4–6.7	0–46	0–200	4–2000	0–96	*	0–7	0–240	0–2400
South Pole[8]	5.4	*	0.16	0.63	0.03	0.16	1.4	1.5	1.3

* Denotes missing values. Single values refer to volume-weighted mean concentrations. All units (except pH) are µeq L^{-1}.

1, Puxbaum *et al.* 1991; 2, Lyons *et al.* 1991; 3, Semkin and Jeffries 1988; 4, Reynolds 1983; 5, Suzuki 1987; 6, Davies *et al.* 1992; 7, Hodgkins *et al.* 1997; 8, Legrand and Delmas 1984.

differential scavenging by vegetation (Kuhn 2001). Snowfall at high altitudes usually contains lower concentrations of solutes than at lower altitudes because the depth of the air column available for aerosol scavenging is smaller. The temporal variability of the chemistry of individual snowfalls is due to the progressive scavenging of the chemical species during the snowfall event, and solute concentrations often decrease in an exponential manner with time (Davies *et al.* 1992).

1.7.1.6 Chemical composition of snow melt

Percolation of melt water through the snow cover causes the chemical composition of both the snow matrix and the meltwaters to change. The concentration and distribution of solutes in the snow–meltwater system is controlled by a variety of physical and biological processes, including the leaching of solute from snow crystals, meltwater–particulate interactions and microbiological activity (Jones 1999; Kuhn 2001).

The leaching of solute from snow grain surfaces by meltwater causes the fractionation of solute into snow melt, often described by a non-dimensional 'concentration' factor, CF_x, $CF_x = {}^xM_m/{}^xM_p$ where ${}^xM_m$ is the concentration of species x in any meltwater fraction and ${}^xM_p$ is the concentration of x in the parent snow *prior* to melt. Values of CF_x during the initial stages of snow melt may reach 50, but more typical values range from 2 to 7. CF_x decreases as the snow ablates, and the final snow melt is much purer that the initial snow pack, with values of CF_x of <0.1 in the final meltwaters. The efficiency of meltwater leaching (i.e. higher values of CF per volume of initial meltwater discharges) depends on the micro- and macroscale distribution of solute in snow grains and meltwater hydrology (Tranter *et al.* 1986; Bales *et al.* 1989; Cragin *et al.* 1996).

The microscale distribution of solute in snow packs prior to melt is influenced by snow metamorphism. Highly concentrated solute is believed to be present as discrete droplets of liquid residing in the boundaries between associated snow grains and as quasi-liquid films on grain surfaces. These liquids contain the bulk of the solute, and are consequently more concentrated than the parent ice crystal by orders of magnitude. The amount of solute able to diffuse from the ice crystal surfaces into meltwater

depends on the diffusion coefficient for the solute, the flow rate, and the length of the flow path (Cragin *et al.* 1996). Deeper snow increases the duration of the snow–meltwater interaction and gives rise to higher snow melt concentrations. For example, Marsh and Webb (1979) report the approximate doubling of initial snow melt concentrations with the doubling of snow depth. Low melt rates promote a more uniform flow of meltwater through the whole snow matrix and solute scavenging is maximized, whereas flow fingering minimizes solute scavenging. The composition of snow melt often varies on a diurnal basis, since solute scavenging is related to the rate of melt. Higher concentrations of solute in meltwater are found in the morning and evening, or during periods of shading, when melt rates are lowest (Bales *et al.* 1993).

The mesoscale distribution of solute in snow cover will also affect the concentration of meltwaters. Discrete snowfalls give rise to snow strata of different compositions. Solute-rich bands also arise from the exclusion of solute from ice lenses formed by the refreezing of meltwater fronts or rain in cold, dry snow. The result of successive diurnal melt–freeze cycles is often to increase the concentration of ions in the first meltwaters issuing from the snow pack. Both laboratory and field experiments have shown that solute-rich layers give rise to more concentrated snow melt (Tranter *et al.* 1988).

Chemical reactions between meltwater and inorganic/organic particles can affect the concentration of solute in the meltwater. The chemical weathering of dust scavenged by snow can take place in the atmosphere or within the snow cover. Many studies have observed the neutralization of snow acidity by carbonaceous dusts from a variety of sources, either of local or remote origin. Organic debris may be found in considerable quantities in some snow covers, for example from canopy litter in boreal forest snow packs. Litter that falls onto snow is relatively unaffected by physical and microbiological activity during the cold accumulation period, while complex reactions occur during the thaw. Field and laboratory studies show that, in general, large amounts of PO_4, K, Mn, Ca, and Mg are discharged from litter-laden snow covers, and there is a decrease in the acidity of the meltwaters. By contrast, interaction between meltwaters and litter often leads to losses of NO_3 and NH_4 in snow cover due to micro-biological activity (Jones 1999; Kuhn 2001).

1.7.2 Biological activity in snow

As indicated earlier, life will only function in liquid water. Within various snow packs and layers there will be times when liquid water occurs, especially during spring and summer melt. The occurrence of areas of red, orange, and green snow on surfaces in the polar regions and on mountains has been noted since early history (Figure 1.28). The first proper recording of the microorganisms responsible for this remarkable colouration was by Bauer in 1819 (quoted in Hoham and Duval 2001). The coloura-tion is due to the so-called snow algae that can reach very high densities of up to 8.6×10^5 cells mL^{-1} of snow water equivalent (Hoham and Duval 2001). The snow algae are not strictly algae, but protozo-ans belonging to the Phytomastigophora (the phytoflagellates).

Organisms that live on the surface of snow are subject to challenging conditions involving high levels of irradiance, including ultraviolet radiation, phases of desiccation, low nutrients, and low tem-peratures. Active phases of the life cycle are restricted to phases when there is liquid water dur-ing spring and summer melt. For most of the year the cells enter resting stages—the hypnoblast. The flagellated vegetative forms only occur when there is liquid water, when they are able to control their position by active movement.

Figure 1.28 Red snow created by snow algae on the Signy Island ice cap, maritime Antarctic. See also Plate 9.

There are many hundreds of species of phytoflag-ellates within 'snow algae' communities. One of the most common, ubiquitous, and most worked on species is *Chlamydomonas nivalis*, but other species are also widely reported, including *Chloromonas nivalis*, *Chloromonas alpina*, and *Raphidonema nivale* (Fogg 1967; Light and Belcher 1968; Hoham and Duval 2001; Remais *et al.* 2005).

Numerous workers have noted that the phyto-flagellates have dust, bacteria, and fungi attached to their cell walls (e.g. Light and Belcher 1968; Hoham and Duval 2001; Remais *et al.* 2005). The question has been posed by Remais *et al.* (2005) that the liv-ing component of this material may be part of some sort of symbiotic association. We know that plank-tonic phytoflagellates exude part of their photosyn-thate, which is subsequently exploited by the bacterioplankton. Could the adhering bacteria and fungi be deriving DOC from their flagellate sub-strate? In return the bacteria and fungi may be pro-viding 'shade' to the flagellate.

Bacteria are a common feature within snow and occur even in the most extreme environments. Bacteria have been recorded in South Pole snow where temperatures can reach –85°C and in summer only rise to –13°C. Bacterial abundances ranged between 200 and 5000 cells mL^{-1} in snow melt in sur-face snow and firm snow. Molecular analysis showed sequences closely resembling the genus *Deinococcus*, a group that is highly resistant to ionizing radiation and desiccation. Perhaps even more remarkable is that the bacterial community displayed measurable growth as manifested by the uptake of radiolabelled thymidine and leucine. Over 2–18 hours radioactive counts in the macromolecular fraction increased by 1.5 to 5 times the counts at the beginning of the incu-bation at –12°C to –17°C (Carpenter *et al.* 2000). It is likely that the bacteria that colonize the snow are of aeolian origin.

Molecular biology not only allows us to see which bacterial taxa occur in an environment, but also provides a valuable ecological tool to follow temporal changes in the succession of species. In the Tateyama Mountains in Japan, one of the high-est snowfall regions in the world, there were marked seasonal differences in the biomass and species dominance. Three species were found in snow during the melting season of which one,

Cryobacterium psychrophilum, was a cold-adapted species (psychrophilic), while the other two taxa were psychrotrophic (with maximum growth rates above 15°C). These were *Variovorax paradoxus* and *Janthinobacterium vividum*. It is not uncommon for species of bacteria in extremely cold environments to be psychrotrophic; while they can sustain growth at low temperatures, their maximum growth rates occur above 15°C. In Antarctic saline lakes, for example, the majority of bacterial species are psychrotrophic. In the Tateyama Mountains *C. psychrophilum* and *J. lividum* increased their biomass significantly between March and April, whereas *V. paradoxus* was slow to grow during the early melt season but continued to increase its growth from June to October (Segawa *et al.* 2005). These differences are probably attributable to differences in the nutrient requirements and tolerances of the bacterial taxa. We know relatively little about the growth requirements of bacteria from these extreme environments.

There are no reports of heterotrophic flagellates from snow, but this does not mean they are not present, it is simply that no-one has looked for them. Where there are bacteria, their predators are also likely to occur. The genus *Ochromonas* has been reported from snow (Fukushima 1963; Fogg 1967) and this phytoflagellate is a well-documented mixotroph in aquatic environments. In a nutrient-limited environment like snow it is likely that it exploits its heterotrophic capability. The community dynamics of snow are very simple with an extremely truncated food web.

CHAPTER 2

Snow

2.1 Snow as an environment

Snow cover offers a physical habitat for life, acts as a reservoir of nutrients, and is a mantle covering soil and vegetation (Gray and Male 1981; Jones 1999). It is important to note that anthropogenic activity from agriculture and industry has impacted on both the chemical loading and the biogeochemical cycling within snow cover. Anthropogenic activity has changed factors such as, for example, the pollutant load and pH of snow cover (Davies *et al.* 1991) and the concentration of biogenic compounds that exchange between snow cover and the atmosphere (Grannas *et al.* 2007). Organic materials of both natural and anthropogenic sources (Stibal *et al.* 2008; Petters *et al.* 2009; Shaw *et al.* 2010) are adsorbed onto small particles in the atmosphere (so-called aerosols), such as soot and mineral grains (Wiacek and Peter 2009). Some organic compounds condense to form organic aerosols. Snow is a more efficient scavenger for aerosols than rain, because of its higher surface area to weight ratio (Davies *et al.* 1984), and hence snow fall effectively sweeps clean the air of particulates. Snow packs can accumulate significant amounts of particulates and solutes from atmospheric deposition, particularly if the air through which it falls is laden with pollution (Tranter *et al.* 1988). Montaine snow packs accumulate both wet (snow, rain, fog) and dry (particulates, gases) atmospheric depositions, which may be held in storage until spring melt, when they are released to the melt runoff and the underlying vegetation. At Niwot Ridge in Colorado, for example, as much as 80% of water input and 50% of inorganic nitrogen deposition can be stored in the snow pack and released at melt (Williams *et al.* 2009).

The results of anthropogenic activities are often carried long distances from their origin and can sometimes produce very clear evidence of pollu-

tion. For example, the High Arctic (Svalbard) was polluted in the spring of 2006 by smoke derived from agricultural fires in eastern Europe (Stohl *et al.* 2007). Biomass burning is a common practice in Russia, Belarus, and the Ukraine and often the fires get out of control, spreading from the crop fields to neighbouring heath and forests. The practice is banned in the European Union. Concentrations of halocarbons, CO_2, and CO derived from the distant fires was the highest ever recorded in Svalbard in an unusually warm spring. The snow on the glaciers was discoloured and consequently snow albedo was reduced. Analysis of snow samples revealed elevated levels of potassium, sulphate, nitrate, and ammonium derived from smoke aerosols. Studies of the long-range transport of nitrogen have shown a doubling of nitrate in firn ice cores taken from the Greenland ice cap since the industrial revolution (Laj *et al.* 1992; Fischer *et al.* 1998). Thus the pollution generated by industry and agriculture is transported to pristine environments where it is deposited. During winter this material accumulates in the snow pack where it undoubtedly supports the growth of microorganism that live in and on the snow surface (snow algae, bacteria). However, at spring melt it provides a pulse of nutrients into the tundra (Hodson *et al.* 2005).

A detailed description of snow structure and formation is given in Section 1.7. Snow has its own biota, but it also strongly influences the habitats that it covers in a variety of ways, as outlined below in Section 2.3.

2.2 Life on and in snow

2.2.1 Snow algae

As described in Section 1.7.2, snow algae are a conspicuous feature of snow (see Figure 1.28). 'Snow algae' are flagellated protozoans that contain photosynthetic pigments (phytoflagellates) and are thus autotrophic, but there are some species amongst them that may be mixotrophic in the snow habitat, for example *Ochromonas* sp.

One of the earliest physiological studies on snow phytoflagellates was conducted by Fogg (1967) on Signy Island in the South Orkneys (Antarctica). He found red and green coloured snow on well-drained firn snow, especially at the margins where the snow had been reduced to a thin layer by abalation. Firn snow is old snow on glaciers or ice caps that is granular and compact, but not yet converted to ice. He examined 90 snow samples and found no consistent differences between red and green snow samples. *Chlamydomonas nivalis*, *Raphidonema nivale*, and *Ochromonas smithii* were common in all samples. The latter species, which is a well-known mixotrophic phytoflagellate, has also been reported from snow fields in Japan (Fukushima 1963). Fogg (1967) measured CO_2 fixation by the phytoflagellates and reported rates of between 0.000041 and 0.000864 $Cm g^{-1} mm^{-3}$ cell material h^{-1}. There was no obvious difference in the ability of the different species to fix carbon. The doubling time of cells based on these rates is around 23 days, which is very slow when compared to temperate species. He also looked at how temperature (between 0°C and 15°C) affected the rate of photosynthesis and found higher rates at higher temperatures. Temperature is not usually limiting to photosynthesis and he attributed the low temperature rates to the effects of desiccation.

Fogg's work used the incorporation of ^{14}C sodium bicarbonate as a measure of carbon fixation during photosynthesis. These days we have access to much more sophisticated methods for measuring photosynthesis including electrodes that measure oxygen evolution. A recent study using this technology confirmed Fogg's findings (Remais *et al.* 2005). There was no decrease in net photosynthesis at higher temperatures suggesting that 'snow algae' are not strictly cryptophytic, or adapted to low temperatures. However, some workers who have cultured these protists have found that they do not survive for long periods at temperatures above 4°C (Hoham 1975). Photoinhibition occurs in many photosynthetic organisms at high irradiances, however *C. nivalis* showed no photoinhibition at irradiances up to 1800 μmolm^{-2}s^{-1} (Remais *et al.* 2005), showing that it is well adapted to an environment subject to high rates of surface photosynthetically active radiation (PAR). These phytoflagellates contain secondary carotenoids, astaxanthin and its fatty ester derivatives, that give the cells their typical red colouration and gave rise to the description of 'red snow'. Such pigments protect the cells from the effects of high irradiation and mediate against photoinhibition.

Snow algae occupy a habitat where they are subjected to high levels of solar radiation, not only PAR but also wavelengths in the ultraviolet (UV) range, which are damaging to biological systems. Phytophenolic compounds play a role in acting as antioxidants in cells. They scavenge singlet oxygen and free radicals—a fact that is well documented for higher plants, but less well so for photosynthetic flagellated protozoans or 'snow algae'. Many natural or synthetic compounds act as antioxidants by trapping singlet oxygen or radicals, thereby terminating chains of reactions. Cell membranes, DNA, and proteins are the main target molecules of free radicals and singlet oxygen. Phenols are particularly effective antioxidants for polyunsaturated fatty acids because they can transfer a hydrogen atom to lipid peroxyl radicals (Foti *et al.* 1994).

When *C. nivalis* was exposed to UV-A (365 nm) for five days and UV-C (254 nm) for seven days, the total phenolic content of the cells changed. In the UV-A treatment the effect was less marked (5–12%) compared with UV-C (12–24%). Antioxidant protection was stimulated (Figure 2.1), but the analysis did not distinguish quantitatively between phenolic compounds, carotenoids, or other known antioxidants. There is no appreciable difference between UV-A and UV-C in terms of the antioxidant protection factor (Duval *et al.* 2000).

Another major challenge faced by snow algae is exposure to freezing conditions. Once snow algae make their appearance they may subsequently experience a heavy freeze. Some species are more readily able to withstand this stress than others. One way of determining the temperature tolerances of individual species is to grow pure cultures at a range of temperatures and to characterize their growth responses. Three species, *Raphidonema*

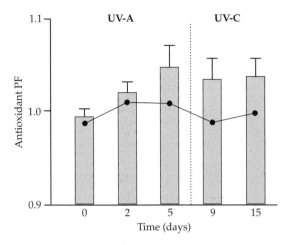

Figure 2.1 Antioxidant protection stimulated by ultraviolet (UV) light. Although antioxidant protection was stimulated by exposure to UV wavelengths, there was significant difference between UV-A and UV-C. Redrawn from Duval *et al.* (2000).

nivale, Chloromonas pichinchae, and *Cylindrocystis brébissonii*, were grown at temperatures between 1°C and 20°C (Table 2.1). The optimum temperatures for growth were 5°C for *R. nivale*, 1°C for *C. pichinchae*, and 10°C for *C. brébissonii*. The latter species had the lowest specific growth rate (μ) of 0.022 day^{-1}, while *R. nivale* had a specific growth rate of 0.116 day^{-1}, and *C. pichinchae* 0.109 day^{-1}. Thus *R. nivale* showed the best adaptation to low temperatures (Hoham 1975). *Chlamydomonas nivalis*, a very common species of snow algae, did not grown in defined media for any length of time; however, it undergoes binary fission (asexual reproduction) at temperatures from 0°C to 2°C which suggests an adaptation to low

Table 2.1 Growth measured as cells mL^{-1} after 32 days at different temperatures for *Cylindrocystis brébissonii* and *Chloromonas pichinchae* and after 30 days for *Raphidonema nivale*, with a photoperiod of 16 hours light and eight hours dark.

	Temperature (°C)				
	1	**5**	**10**	**15**	**20**
Raphidonema nivale	125.1	275.9	106.3	84.0	*
Chloromonas pichinchae	19.9	17.4	1.7	**	**
Cylindrocystis brébissonii	1.3	1.6	3.1	2.9	2.2

* Did not survive beyond day 22;
** did not survive beyond day 28.
Data from Hoham (1975).

temperatures and an optimum growth temperature close to this.

2.2.2 Bacteria in snow

Given that snow in many parts of the world is subject to atmospheric deposition of nitrogen, phosphorous, and carbon, one might reasonably expect to find bacteria growing in snow that exploit these nutrients. Bacteria have been reported as actively growing in snow at the South Pole in summer. Counts of cells in melted snow reached 5000 cells mL^{-1}. Moreover, measurements of DNA and protein synthesis, as indicated by the uptake of radiolabelled thymidine and leucine, indicate that at least some of these cells were actively growing at ambient sub-zero temperatures (Carpenter *et al.* 2000). These data have been questioned, however, by Warren and Hudson (2003) who argued that for a variety of reasons there would be insufficient liquid water to support functional bacteria. They did not dispute the presence of bacteria but argued that these may have been carried by the wind from elsewhere. Carpenter and Capone (2003) were able to provide sound arguments to refute the criticism.

By contrast, there is little doubt that microbial activity and the utilization of dissolved nitrogen, such as NO_3^- and NH_4^+, occurs in the warmer and wetter snow packs that are characteristic of the boreal forests of the northern hemisphere in the spring (Jones 1999). In general, careful mass balance studies of these species show a loss when the snow is melting and wet. However, dry deposition of these species from the atmosphere confounds this simple assertion on occasion, as does photolysis (Grannas *et al.* 2007). In general, the total nitrogen content of the snow pack is small compared to that of the surface soil layers beneath the snow pack. However, the nitrogen contained in snow melt is usually very labile, whereas some of the nitrogen contained in the soil water may be in more refractory, dissolved organic phases (Jones 1999). Warmer and wetter snow packs are also found on glaciers in the maritime Antarctic, such as Signy Island in the South Orkney Islands. Here, NH_4^+ and phosphorus are taken up by microbial activity in the snow pack during periods of snow melt. An additional stimulus to nutrient cycling comes from penguin and seal excretia, and additional NO_3^- may be derived from the dissolution of rock debris on the margins of the snow pack (Hodson 2006). Large mammals, such as caribou, may also add significant quantities of nutrient locally to boreal forest snow cover via excretion (Jones 1991).

In the Arctic at Ny Ålesund in Spitzbergen, where there may be significant atmospheric deposition, bacteria numbers were much higher than at the South Pole, at around $6 \times 10^4 \, mL^{-1}$ throughout the 60cm deep snow pack (Amato *et al.* 2007). In the European Alps, bacterial concentrations in snow were lower and ranged from 1.1×10^4 to $2.0 \times 10^4 \, mL^{-1}$ (Sattler *et al.* 2001; Bauer *et al.* 2002). However, the higher figure was for snow mixed with graupel, a German term for small hail or soft hail, which forms when supercooled droplets of water condense on a snow flake. Sattler *et al.* (2001) recorded 9.2×10^3 bacteria mL^{-1} in graupel. In contrast, much higher figures were recorded in the Tateyama Mountains in Japan, an area that has one of the highest snow-falls in the world. Here bacterial densities ranged from 6.0×10^3 up to $2.3 \times 10^5 \, mL^{-1}$ (Segawa *et al.* 2005). Clearly there are very considerable variations in the concentrations of bacteria found in snow, particularly in mountain areas at lower latitudes. As yet we lack the data to determine the major factors that control the biomass of bacteria communities, but the availability of liquid water, a carbon source, and the macronutrients nitrogen and phosphorus are likely to be prime factors.

Molecular analysis of bacterial strains isolated from Arctic snow revealed members of the Alphaproteo-bacteria, Betaproteobacteria, Gammaproteobacteria, Firmicutes, and Actinobacteria (Amato *et al.* 2007). An analysis of the whole snow community in samples from the Tateyama Mountains in Japan revealed members of the same groups (Segawa *et al.* 2005). This latter study followed seasonal changes in the biomass some of the dominant taxa in the community (Figure 2.2). As the graph shows there was a rapid increase in biomass from March to April, with a maximum in August. This follows the progression of the melt season.

Mass balance studies of NO_3^- and NH_4^+ of snow pack and runoff from glaciers in Svalbard show that NH_4^+ is apparently lost from the surface snow, whereas NO_3^- seems to increase in runoff (Hodson *et al.* 2005). It would be tempting to sug-

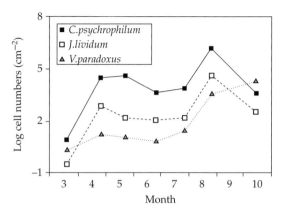

Figure 2.2 Seasonal changes in numbers of the major bacteria (*Cryobacterium psychrophilum*, *Varovorax paradoxus*, and *Janthinobacterium lividum*) in the surface snow in the Tateyama Mountains of Japan from March to October. Redrawn from Segawa *et al.* (2005).

gest that the NH_4^+ is converted to NO_3^- by biological activity within the snow pack, but a more forensic interpretation of the seasonality of the changing mass balances suggests that whereas NH_4^+ is fixed as organic matter on the glacier surface, NO_3^- seems to be generated in subglacial environments. The locus of NH_4^+ fixation may well be wet snow, but there is compelling evidence to suggest that cryoconite holes are a more likely locus (Hodson *et al.* 2005). We examine this idea further in Section 3.2.3.

Bacteria that live in very low temperature environments have evolved physiological and biochemical adaptations that enable them to survive freezing and maintain function. Most of the research on such adaptations has been conducted on species isolated from polar lakes, but the findings are equally applicable to species that inhabit snow and ice. Major interest has been focused on low temperatures enzymes, cold shock induction, and ice-active substances. These biochemical findings have been of considerable interest to industry because of their potential applications in the food industry and pharmaceutical and industrial processes involving enzymes. Ice-active substances such as a cold shock inducible peptidyl-prolyl *cis*-transferase (Ideno *et al.* 2001) and a hyperactive Ca^{2+}-dependent antifreeze protein (Gilbert *et al.* 2005) have been identified. Antifreeze proteins allow the organism to control the size of the ice crystals and thereby limit damage to the cell. Enzymes with a low temperature optima have been isolated, for example a β-galactosidase from an Antarctic soil bacterium (Coker *et al.* 2003), a cold active alkaline phosphatase (Dhaked *et al.* 2005), and a low temperature lipase (Yang *et al.* 2004). A major challenge as freezing occurs is maintaining the fluidity of the cell membrane. The cell membranes of bacteria develop a homeoviscous state as temperatures decrease, involving an increase in the level of unsaturated fatty acids, a decrease in the fatty acid chain length, and increasing branching of the chains (Sinensky 1974; Kumar *et al.* 2002).

2.3 Impact of snow on environments it covers seasonally

2.3.1 Activity under the snow

While snow itself offers a habitat for microbial colonization and activity, as we have seen in the preceding sections, snow can also have a profound impact on the ecosystems its covers. In alpine regions the landscape is covered each year by snow to varying depths. It was assumed that during this phase biological activity in the underlying soil ceased. Recent research indicates that this is not the case at all, and that under the insulating layer of snow microbial activity in the soil and at the soil–snow interface can be significant. A number of studies applying a range of techniques have demonstrated an active microbial community under snow in winter and into spring (Monson *et al.* 2006; Bowling *et al.* 2009; Lipson *et al.* 2009a, 2009b). One of the main approaches adopted is to measure fluxes of CO_2 from the soil through the snow into the atmosphere. The CO_2 is released during respiration by soil microorganisms. The presence of enhanced CO_2 under snow related to biological activity has been recognized for four decades (Kelley *et al.* 1968). Modern analytical techniques and the use of carbon stable isotope chemistry, together with advances in microbiological and molecular techniques, have allowed us to gain a very detailed picture of under-snow processes.

Snow acts as an insulating layer and consequently its thickness will have an effect on soil temperature, which in turn will affect microbial community metabolism. Soil temperature correlates with net ecosystem CO_2 exchange and with snow water content (Figures 2.3 and 2.4) (Monson *et al.* 2006). Analysis of a six-year data set showed that in years where there was a reduced snow pack there were lower rates of soil respiration. Despite the low temperature of the soil beneath the snow, there was significant respiration and release of CO_2 (Figure 2.3), but this process is sensitive to small changes in soil temperature relating to snow thickness.

It appears that the heat produced by microbial metabolism can produce isolated pockets of activity in what are very cold soils. Measurements have

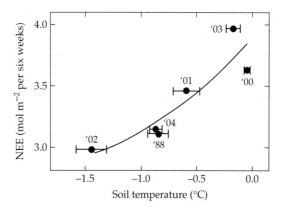

Figure 2.3 The responses of net ecosystem CO_2 exchange (NEE) to average daily soil temperature. Redrawn from Monson *et al.* (2006).

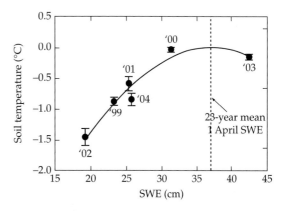

Figure 2.4 The relationship between snow water equivalent (SWE) (1 April) and average daily soil temperature. Data from 1 April until 15 April. Redrawn from Monson *et al.* (2006).

shown that microbial biomass can be higher in winter than in summer (Bowling *et al.* 2009). Furthermore subnivian soil microbial communities can have higher growth rates but lower growth yields than summer and autumn communities in exposed soils (Lipson *et al.* 2009a). This means that the winter community has higher biomass-specific respiration rates than summer and autumn communities. The cost of maintenance is undoubtedly higher at low temperatures so that less energy is available to partition into growth. Bacteria had higher growth rates and lower yields than fungi indicating that they may have a more significant role in determining CO_2 fluxes from the soil in winter.

The fluxes of CO_2 from snow-covered soil appear to make a significant contribution to annual CO_2 efflux at alpine and subalpine sites. At a subalpine site in Colorado (Niwot Ridge) CO_2 fluxes of 0.71 and 0.86 $\mu molm^{-2}s^{-1}$ were recorded in 2006 and 2007, respectively (Lipson *et al.* 2009b). These are among the highest values reported for snow-covered systems and at Niwot Ridge are equivalent to around 30% of annual CO_2 efflux. In general, CO_2 flux increased in winter as soil moisture increased. Lipson *et al.* (2009b) constructed a model related to different snow cover zones. They showed that as snow depth and duration increased, the factor controlling CO_2 flux shifts from freeze–thaw cycles to soil temperature to soil moisture to carbon availability. Quite clearly this has implications as climate changes. Interannual variability in snow cover (depth and extent) is likely to result in different patterns of CO_2 flux from seasonally snow-covered soils.

Molecular analysis has shown that the microbial communities in soil beneath snow differ from those that occur in the exposed soil in summer and autumn. 16S ribosomal RNA clone libraries from winter and summer showed that 38% of the winter sequences related to *Janthinobacterium*, while in summer 17% of the sequences were closely related to *Burkholderia*. The overall picture of DNA libraries differed significantly between summer and winter (Monson *et al.* 2006). Different taxonomic makeup implies differing physiological capacity over the annual cycle. The summer community was unable to grow below 4°C while the winter community was able to grow exponentially at 0°C and their growth rates responded quickly to increased temperature.

While temperature and the availability of soil moisture are prime factors in supporting growth, the supply of a suitable carbon substrate is also crucial. Increased soil carbon correlates well with CO_2 flux (Figure 2.5). As soil carbon content increases so does CO_2 flux, reflecting a larger, more physiologically active microbial community. It is relatively easy to determine experimentally if a community of microorganisms is carbon limited, by adding carbon sources either in the field or *in situ* and then measuring the resultant carbon flux. *In situ* studies at the Upper Snake River and Deer Creek in Colorado, where additions of labile carbon (glucose) were added to snow-covered soils through snow pits, produced an immediate increase in subnivial CO_2 concentration (Figure 2.6) (Brooks *et al.* 2004).

The soil surface–snow interface also appears to be an important habitat for microbial colonization in winter in conifer and tundra systems. Mats of snow molds or fungi are commonly observed on the soil and litter surface as the snow packs recede in spring. The molds disappear as the snow is lost (Schmidt *et al.* 2009). Molds cultured from the natural environment at 3050m altitude in a pine forest in Colorado were members of Mortierellales and Mucorales, which are subdivisions of Zygomycota. Growth experiments in the laboratory showed

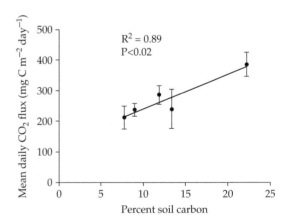

Figure 2.5 The relationship between overwinter CO_2 flux and soil carbon content along five 1km transects in the Summit country of Colorado. Measurements were made biweekly between January and May at 10 locations along each transect. From Brooks *et al.* (2004), with permission of John Wiley and Sons.

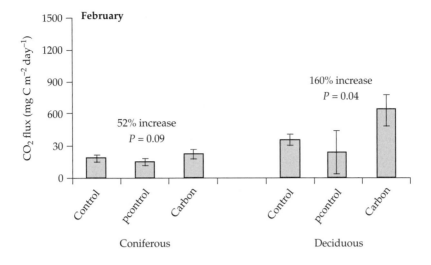

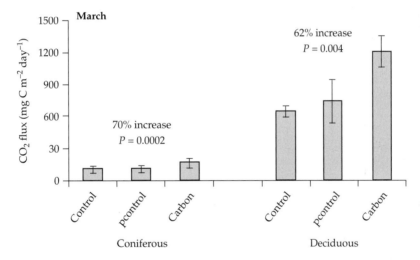

Figure 2.6 The response of subnivial soil respiration to additions of carbon. The top panel shows the response after 24 hours in February and the bottom panel after 30 days in March. Control, natural snow pack; pccontrol, where snow was removed; carbon, where carbon (glucose) was added. Data for both coniferous and deciduous habitats are shown. From Brooks *et al.* (2004), with permission of John Wiley and Sons.

strong exponential growth at sub-zero temperatures of –2°C and –3°C (Figure 2.7). These snow molds have unusually high Q_{10} values as shown in Table 2.2. Q_{10} values are useful in characterizing the magnitude of increase in metabolic rates as temperatures rise. They have been used to give an indication of the temperatures tolerances of a species. These very high Q_{10} values show a strong response to very small changes in temperature by snow fungi that are clearly extremely well adapted to low temperatures.

Table 2.2 Q_{10} values for snow molds between –2°C and –3°C and –3°C and 3.8°C derived from the equation $Q_{10} = (r_2/r_1)^{10/(T2-T1)}$ where r_2 and r_1 represent the respiration rates of the molds at temperatures T_2 and T_1 respectively.

	Isolate		
	Mortierellales	*Mucorales* strain a	*Mucorales* strain b
Q_{10} –2°C to –3°C	330	22.5	277
Q_{10} –3°C to 3.8°C	1.2	4.7	2.0

Data from Schmidt *et al.* (2009).

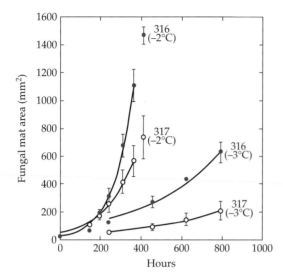

Figure 2.7 Exponential growth rates of snow mold isolates 316 and 317 at −3°C and −2°C. From Schmidt *et al.* (2009), with kind permission of Springer-Science+Business Media B V.

2.3.2 Impact of the release of accumulated nutrients in the snow pack at spring melt

As indicated in Section 2.1 pollutants are transported from sites of industrial and agricultural activity to the Arctic, where they are deposited. During the phase of annual snow cover these chemicals accumulate in the snow pack and are released as a large pulse in spring. The question is what happens to these chemicals, especially species of nitrogen, at a time when the underlying vegetation is becoming metabolically active in the short Arctic summer. One way of answering this question is to apply N^{15}-labelled ammonium and nitrate to the snow pack to mimic the spring melt pulse at concentrations that equate to the annual inorganic N deposition and then to trace its fate in the soil and vegetation. An investigation using this approach was undertaken in Spitsbergen in the Svalbard Archipelago at 79°N (Tye *et al.* 2005).

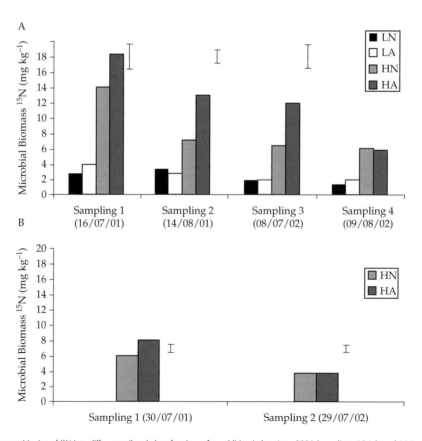

Figure 2.8 Mean partitioning of ^{15}N into different soil and plant fractions after addition in late June 2001 (samplings 16 July and 14 August) at two sites (A) Brandalpynten and (B) Ny-London. LA, 0.1 g m^{-2} NH$_4$-N; LN, 0.1 g m^{-2} NO$_3$-N; HA, 0.5 g m^{-2} NH$_4$-N; HN, 0.5 g m^{-2} NO^3-N. From Tye et al. (2005), with permission of John Wiley and Sons.

Two tundra sites were selected along the Kongsfjorden, one north facing and dominated by moss with low-growing dwarf willow *Salix polaris* and *Saxifraga oppositifolia* and the other south facing and dominated by bryophyte, graminoid, and lichen species. Labelled ^{15}N (99 atom%) was applied to plots as either $Na^{15}NO_3$ or $^{15}N\text{-}H_4Cl$ at rates of either 0.1 or $0.5\,g\,N\,m^{-2}$, which equates to the yearly inorganic nitrogen deposition on Spitzbergen of $0.1\,g\,N\,m^{-2}$ (Woodin 1997). The initial partitioning of ^{15}N in the first summer showed that around 60% of the applied nitrogen was recovered from the soil, litter, and plants irrespective of the rate of applica-

tion. This showed that the pulse of nitrogen is rapidly immobilized into organic forms, particularly into microbial biomass, bryophytes and fine humus (Figures 2.8 and 2.9). In the ensuing two to three seasons there was very little change in the levels of ^{15}N recovered indicating that there is high conservation of nitrogen, as shown in Figure 2.9 for microbial mass. One of the clearest transfers of nitrogen was from the microbial biomass into stable forms of humus. The tundra was very wet as the snow melted, with a microbial community that resembled that of aquatic environments with diatoms, desmids, rotifers, and the cyanobacterium *Nostoc*. As it dried

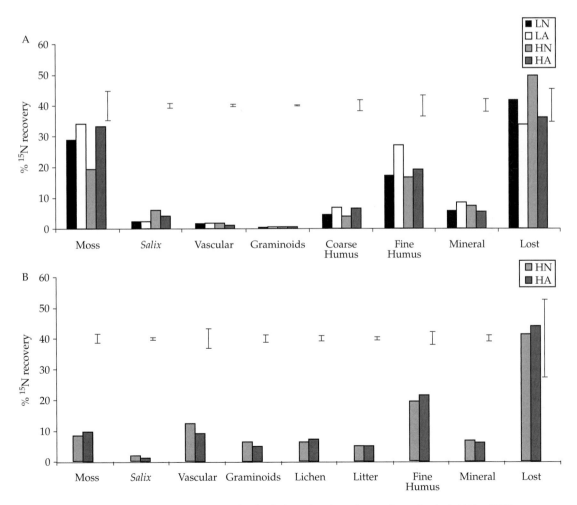

Figure 2.9 Partitioning of added ^{15}N into microbial biomass found in the humus layer (0–3 cm) at samplings undertaken in 2001 and 2002 at (A) Brandalpynten and (B) Ny-London. LA, $0.1\,g\,m^{-2}\,NH_4\text{-}N$; LN, $0.1\,g\,m^{-2}\,NO_3\text{-}N$; HA, $0.5\,g\,m^{-2}\,NH_4\text{-}N$; HN, $0.5\,g\,m^{-2}\,NO^3\text{-}N$. From Tye et al. (2005), with permission of John Wiley and Sons.

out these elements disappeared, though the diatoms remained. As Figure 2.8 shows, 40% of the applied nitrogen was lost in runoff from the tundra to the fjord. Thus deposited nitrogen and other ions have the potential to enrich production in the surrounding marine environment in spring and summer.

2.3.3 Variations in snow depth

In southerly tundra boreal forest ecosystems, variation in snow depth can have an impact on vegetation structure by contributing to competition between vascular and non-vascular plants such as lichens. Shallow snow less than 30cm in depth favours non-vascular plants as it provides protection from wind abrasion and desiccation. In contrast trees benefit from a thick snow cover that provides insulation. Trees serve to trap snow in drifts. Mean snow cover depth is an important factor in determining the height of the most productive plant growth. Foliage that is snow covered remains healthy, while that exposed above the snow gets subjected to abrasion. This material accumulates on the snow and contributes to the litter layer at snow melt (Jones 1991).

CHAPTER 3

Ice surface environments

Liquid water needs to be present within snow, glacier ice, or debris for microorganisms to be viable on the surfaces of ice shelves, glaciers, and ice sheets. Since our understanding of snow ecology has been derived largely from the study of non-glacial snow packs, it is the debris-rich habitats on the glacier surface that are by far the most understood components of supraglacial ecosystems. Below a distinction is made between ice shelves and glaciers because the gentle surface slopes of the former allow the debris to form complex mat-like structures that are not widely observed on glaciers. By contrast glaciers and ice sheets have steeper surface gradients so debris tends to be dispersed widely over the ice surface, resulting in the formation of millimetre-scale aggregate particles known as 'cryoconite'. Other habitats upon glaciers include supraglacial streams, lakes, and ice marginal debris, all of which are considered below.

3.1 Ice shelves

3.1.1 Introduction

Ice shelves are a feature of both polar regions, though the majority, and the largest, are in the Antarctic, for which 43 ice shelves are listed. These include the Amery Ice Shelf, the George VI Ice Shelf, the Ronnie Ice Shelf, the Larsen Ice Shelves, and the McMurdo Ice Shelf. Ice shelves in both polar regions have undergone catastrophic breakup in the past, resulting in the sudden disintegration of the ice into the ocean in which it floats. The best known recent events were associated with the Larsen A and B Ice Shelves during 1995 and 2002 respectively (Scambos *et al.* 2003). In the Canadian High Arctic there are a series of remnant ice shelves along the northern coastline of Ellesmere Island (83°N). Prior to the

twentieth century, the northern coast of Ellesmere Island was fringed by thick land-fast ice floating on the sea, forming an extensive ice shelf that developed about 4500 years ago. Based on various expedition data and satellite images, estimates suggest that this ice shelf had an extent of around 9000 km². By 1999 it had shrunk by 90% to a series of ice shelves contained in embayments along the coast. Of these, the largest is the Ward Hunt Ice Shelf. The remnant ice shelves are further fragmenting, probably as a result of climate warming (Vincent *et al.* 2004).

The mass balance of ice shelves is achieved through a variety of processes. While some are fed by snowfall and glacier ice, others represent the long-term accumulation of sea ice or are sustained by basal freezing of underlying waters. Ice is lost from ice shelves by calving, melting, and ablation. In many ice shelves quantities of marine sediment are brought to the surface of the ice by basal freezing and surface ablation and are redistributed by stream flow and wind activity across the ice shelf. The distribution of the sediment on the surface varies considerably, from almost complete sediment cover to a spare distribution of sediment. Where thick sediment is present it acts to insulate the ice and inhibits melting, producing a stable ice surface within which large perennial ponds lined by sediment may develop. Typically, ice shelf surfaces are undulating with elongated lakes and streams that occupy troughs that are aligned in a parallel fashion (Hawes *et al.* 2008) (Figure 3.1).

One of the most extensive surface ablation areas in Antarctica is thought to be the McMurdo Ice Shelf, which is virtually static so that accumulation via freezing of seawater at its base is offset only by ablation from its surface. Sediment material incorporated into basal freezing is transported

Figure 3.1 Ward Hunt Ice Shelf (Canadian Arctic) showing parallel, elongated surface lakes. Photo courtesy of W. F. Vincent. See also Plate 10.

up through the ice shelf following surface ablation, forming a significant debris layer (Debenham 1920). Sediments here are dark and absorb heat, leading to the development of lakes and ponds during summer. These water bodies may be ephemeral where sediments are sparse, or long-lived ice shelf ponds and lakes where sediments are more extensive.

3.1.2 Biology of ice shelf lakes

In the more ephemeral water bodies rapid coloniz-ers appear, such as coccoid chlorophytes and dia-toms, for example *Pinnularia cymatopleura*. In the perennial ponds and lakes more long-lived com-munities become established and persist from year to year. These communities are usually dominated by cyanobacterial mats, with species composition varying in relation to the conductivity of the water. Typically mats are made up from species of *Oscillatoria*, *Phormidium*, *Nostoc*, and *Nodularia*. In salinities greater than 70 mS cm[-1] *Oscillatoria* cf. *priestleyi* dominates Antarctic mats, while *Nodularia* is more common in brackish ponds and lakes. Cyanobacterial mats can build up high biomass; for example in ponds on the McMurdo Ice Shelf, cyanobacterial mats reach concentrations of chloro-phyll *a* as high as 400 mg m[-2] (Hawes *et al.* 2008). Benthic species of diatoms are common among cyanobacterial mats, particularly species of *Navicula*. The water overlying mats may contain a

community of phytoplankton species such as *Ochromonas*, *Chlamydomonas*, and *Chroomonas*, all species that are typical of polar lakes. Planktonic ciliates typical of polar lakes also occur in the plankton where they feed on bacteria and flagel-lates (James *et al.* 1995).

Cyanobacterial mats represent the dominant bio-mass in ice shelf ponds and lakes (Figure 3.2). They are characterized by steep and fluctuating physio-chemical gradients. Among the most important of the parameters are oxygen, sulphide concentration, and light. The Cyanobacteria are unique among the prokaryotes in that they possess the ability to under-take oxygenic, plant-like photosynthesis in which two photosystems, PSII and PSI, are connected in series. Much of the organic matter produced by Cyanobacteria during photosynthesis is eventually decomposed. Among the decomposers, sulphate-reducing bacteria may be important. These bacteria are obligate anaerobes and produce sulphide. The gradients of sulphide and oxygen in mats vary in the sediment as light intensity varies diurnally. Periodically mat-dwelling Cyanobacteria may be exposed to high concentrations of toxic sulphide (Stal 1995). While cyanobacterial mats occur in aquatic ecosystems worldwide, they are a distinct feature of extremely cold environments including surface ice environments, polar streams, and polar lakes.

The dominance of Cyanobacteria in these extreme environments is particularly interesting as they are a group that is generally regarded as having high optima temperatures for growth in excess of 20°C. A study that grew 27 Arctic, sub-Arctic, and Antarctic strains of *Oscillatoriaceae* at 5°C intervals between 5°C and 35°C found that all of the strains had temperature optima for growth between 15°C and 35°C (mean 19.9°C). Quite clearly these strains isolated from polar environments are not psy-chrophiles (with growth optimum temperatures below 15°C), but psychrotrophs able to colonize and survive extremely low temperatures (Tang *et al.* 1997). This is also the case for many heterotrophic bacteria that have been investigated from polar lakes. While some are psychrophiles, the majority studied so far are psychrotrophs. Low temperatures confront organisms with a raft of challenges in addi-tion to the need to function at temperatures close to

Figure 3.2 A cyanobacterial mat on the bottom of the littoral region of an Arctic lake. Photo courtesy of W. F. Vincent. See also Plate 11.

or below freezing. In ice shelf ponds and lakes, seasonal freezing of the water column from top to base requires an ability to withstand freeze–thaw cycles and a tolerance to desiccation. In summer an ability to withstand high levels of irradiance, including ultraviolet (UV) wavelengths that cause severe physiological and genetic damage, is essential. As winter encroaches, ponds, some of which may be saline, freeze up and diminishing liquid water becomes progressively more saline as salts are excluded from the forming ice, requiring organisms to withstand wide changes in salinity. Cyanobacteria possess the physiological and biochemical plasticity to adapt to these environmental challenges, providing a competitive advantage that compensates for not being psychrophilic.

The ability to withstand high irradiances is achieved through the possession of pigments such as scytonemin and carotenoids. Cyanobacterial mats have a clear structure with a zonation of vari-

ous pigments including chlorophyll. This gives a stratified appearance with distinct variations in colour. Active mats trap sediment and bind it into the mat structure. Such structures have given rise to consolidated rock known as stromatolites that are some of the earliest forms of life in the fossil record. In common with diatoms (see Section 1.2.2.3), Cyanobacteria excrete extracellular polymeric substances composed primarily of polysaccharides, of which there are two distinct types: capsular polysaccharides (cps) and extracellular polysaccharides (eps). The former are associated with the capsules and sheaths surrounding the organisms and are regarded as largely recalcitrant. In contrast eps are freely present in the medium and probably act in the gliding movement of the trichomes. They assist in binding sediments and undoubtedly provide a carbon source for heterotrophic bacterial growth (Stal 1995).

Mats in water bodies on the Markham Ice Shelf in the Arctic have provided an example of the zonation of pigments. They were overlain by a thin cohesive surface layer between 100 and 150 μm thick, made up of terrestrial and subaerial palmelloid chlorophytes resembling *Chlorosarcinopsis*, *Pleurastrum*, *Chlorokybus*, and solitary cells of *Bracteococcus*. Some were green while others had an orange-yellow pigmentation. These organisms were held in a matrix of Cyanobacteria composed of *Nostoc*, *Gloeocapsa*, *Phormidium*, *Leptolyngbya*, and diatoms (*Navicula* spp.). This surface layer was underlain by a centimetre-thick layer of flocculent, dark-coloured material in contact with the ice, containing chlorophytes, diatoms, *Gloeocapsa*, and *Leptolynbya*, some being partially decomposed, and heterotrophic bacteria. Within the mat, Cyanobacteria represented between 88% and 89% of the total cell counts (Vincent *et al.* 2004).

Detailed analysis of the mat pigments using high performance liquid chromatography (HPLC) revealed a range of pigments that varied in their concentration throughout the mat profile (Vincent *et al.* 2004) (Table 3.1). A scytonemin-red-like compound was recorded in both layers but was 50% higher in the bottom layer. This was probably a decomposition product decaying slowly in the low temperatures. The surface layer contained high levels of chlorophylls *a* and *b* and chlorophyllide and a

Table 3.1 Pigment concentrations and ratios in the surface and bottom layers of mat communities in the Markham Ice Shelf, High Arctic in summer 2003.

Pigment	Pigment concentration ($\mu g\ cm^{-2}$)		Pigment ratios, surface : bottom
	Surface layer	**Bottom layer**	
Chlorophyll *a*	32.95	17.19	1.92
Chlorophyll *b*	8.17	2.74	2.98
Chlorophyllide	3.42	0.73	4.70
Scytonemin	69.72	99.94	0.70
Scytonemin-red	60.94	169.78	0.36
Total carotenoids	20.12	4.35	4.63

Data from Vincent *et al.* (2004).

variety of carotenoid compounds. The latter play an important role in providing protection from high solar irradiance, including UV wavelengths. As well as having protective pigments, there is evidence that some Cyanobacteria in Antarctic mats have deep living trichomes that migrate upwards to the mat surface in response to decreasing surface irradiance, thereby avoiding photochemical damage (Vincent *et al.* 1993).

The Cyanobacteria possess the ability to fix atmospheric nitrogen and thus have an advantage in environments where inorganic nitrogen may be limiting to growth. Nitrogen fixation also plays an important role in the nitrogen budgets of many ecosystems. Heterocyst-bearing species are capable of fixing nitrogen under fully oxic conditions in the light, while non-heterocystous species can fix atmospheric nitrogen under oxic conditions usually in the dark. Heterocystous species include *Nostoc*, *Anabaena*, and *Nodularia*, while non-heterocystous forms include *Phormidium*, *Oscillatoria*, and *Lyngbya*. Nitrogen fixation is mediated by the enzyme nitrogenase. Dinitrogen is inert because of the strength of its N-N triple bond, thus breaking one of the nitrogen atoms from another requires all three of the chemical bonds to be broken. Nitrogenase therefore has a high requirement for energy. The activity of this enzyme in Cyanobacteria and other nitrogen fixers is assessed by acetylene reduction activity (ARA).

Nitrogen fixation, which occurs in anoxic conditions, must therefore be separated from oxygenic photosynthesis within the organism. Heterocystous cynaobacterial species have evolved specific cells for nitrogen fixation—the heterocysts (Figure 3.3). These cells do not undertake photosynthesis. They possess a thickened cell wall that resists the diffusion of gases. Within the heterocyst, respiratory activity renders the cell virtually anoxic. Thus the heterocystous bacteria have functionally separated oxygenic photosynthesis from nitrogen fixation, and the two processes can operate in tandem.

Cyanobacterial mats are very well developed in the many ponds and lakes on the McMurdo Ice Shelf. Within mats *Oscillatoria* (a non-heterocystous genus) accounted for 70% of the total cell count. Hetercyst-bearing species such as *Nodularia* and *Nostoc* did not contribute a large portion of the mat communities. During 24-hour summer daylight nitrogenase activity followed a diurnal cycle, reaching its maximum at midday and falling to

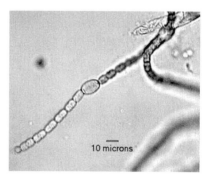

Figure 3.3 The cyanobacterium *Nostoc* showing a heterocyst (the larger cell in the filament). Photo courtesy of Roger Burks and Mark Schneegurt and Cyanosite (www-cyanosite.bio.purdue.edu).

Table 3.2 A nitrogen budget for the water column of a McMurdo Ice Shelf pond.

Process	Flux (mg m^{-2} day^{-1})	Percentage of inflow
In		
Precipitation	0.2	1
Nitrogen fixation	12.8	67
Recycled nitrogen	6.0	32
Total in	19.0	100
Out		
Denitrification	0.4	2
Phytoplankton uptake	3.8	20
Benthic uptake	4.8	26
Storage*	10.0	52
Total out	19.0	100

* By difference.
Data from Hawes *et al.* (2008).

its lowest in the early hours of the morning. Nitrogen fixation was confined to the upper layers of the mats and was achieved by heterocystous species. Estimates suggest that in the McMurdo Ice Shelf ponds mat rates of nitrogen fixation amount to 12.8 mg m^{-2} day^{-1}, accounting for 67% of total nitrogen inputs (Fernández-Valiente *et al.* 2001; Hawes *et al.* 2008). A nitrogen budget for a typical McMurdo Ice Shelf pond developed by Hawes *et al.* (2008) (Table 3.2) estimates that most of the remaining 33% of the nitrogen budget is derived from nitrogen released from the sediments as recycled ammonium/nitrogen, with only 1% derived from snow and ice melt. Much of the nitrogen (98%) is retained in the system, mainly incorporated into the biota. The retention of dissolved organic nitrogen (DON) is high in these ponds and can exceed 13 g m^{-3}, but it appears to be very refractory and consequently accumulates.

The biogeography of polar Cyanobacteria is particularly interesting. Based on morphology and traditional taxonomic analysis there did not appear to be any significant endemism among the Cyanobacteria that form mats in Antarctic water bodies, but when subjected to molecular analysis there was clear evidence of endemism (Taton *et al.* 2006). Investigations of Cyanobacteria from a range of habitats in the High Arctic, including ice shelves, revealed that a number of genera previously identified as endemic to Antarctica (*Phormidium priestleyi*, *Leptolyngbya frigida*, and *Leptolyngbya antarctica*) were 99.0% similar to the sequences of Canadian High Arctic Cyanobacteria (Jungblut *et al.* 2010). Thus there appear to be cold habitat-specific genera common to the polar regions that are absent from lower latitude climatic zones.

Despite the high biomass contributed by Antarctic mats, as indicted by areal chlorophyll *a* (up to 400 µg cm^{-2}), the rates of photosynthesis are low (Table 3.3). The biomass in Antarctic mats found in ice shelf ponds is often comparable to that seen in temperate water bodies. Arctic ice shelf ponds have lower biomass, for example on the Ward Hunt Ice Shelf chlorophyll *a* ranged from 5.63 to 30.58 µg cm^{-2} (Mueller *et al.* 2005) and in Markham Ice Shelf ponds up to 32.95 µg cm^{-2} in the upper layer (Vincent *et al.* 2004). However, mats in Arctic ponds are more productive and have reasonably high assimilation numbers (Table 3.3). Given the paucity of photosynthesis data caution needs to be exercised when making comparisons.

The majority of studies pertain to the summer when cyanobacterial mats are physiologically

Table 3.3 Rates of photosynthesis in microbial mats in ice shelf ponds during summer.

Site	Photosynthetic rate (mg C m^{-2} h^{-1})	Assimilation number (g C (g Chl a)$^{-1}$ h^{-1})
Ponds on Ward Hunt Shelf 83°N[1]	27.3–105	0.059–0.17
McMurdo Ice Shelf 78°S (six ponds of varying salinity)[2]	14.7–42.6	0.003–0.043
McMurdo Ice Shelf 78°S (six ponds of varying salinity)[3]	10–40	0.048–0.275

1, Mueller *et al.* (2005); 2, Howard-Williams *et al.* (1989); 3, Vincent *et al.* (1993).

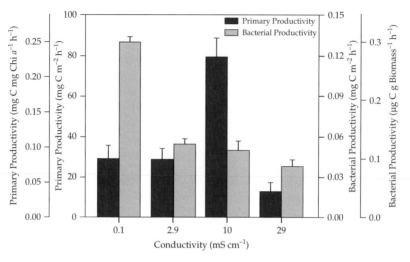

Figure 3.4 The effect of increasing conductivity on bacterial production (grey columns) and primary production (black columns) in a microbial mat dominated by Cyanobacteria on the Ward Hunt Ice Shelf (Canadian High Arctic). Redrawn from Mueller *et al.* (2005).

functional. There is only limited information on how mats respond to the environmental stresses that occur as winter encroaches. Hawes *et al.* (1999) studied two ponds on the Ross Ice Shelf for a year from January to January. The lower waters of the deeper pond remained unfrozen until June, but the water increased in conductivity as the liquid phase diminished. This pond increased in conductivity to $60\,mS\,cm^{-1}$ before freeze-up. Mats were able to remain photosynthetically active as conductivity increased. Experimental data showed that mats can function in conductivities up to $80\,mS\,cm^{-1}$ and can acclimate to reducing irradiance levels. Arctic mats also show a good tolerance to changes in conductivity (Figure 3.4). Primary production increased progressively as conductivity increased from 0.1 to $10\,mS\,cm^{-1}$, and then declined as conductivity increased up to $29\,mS\,cm^{-1}$ (Mueller *et al.* 2005). Interestingly, heterotrophic bacteria showed a progressive decline in productivity in response to increased conductivity (Figure 3.4).

Heterotrophic bacteria have been less well studied in mats. The limited information suggests that they are well adapted. An Arctic community achieved its highest production at lower temperatures. Bacterial production ranged from 0.037 to $0.21\,mg\,C\,m^{-2}h^{-1}$ (mean 0.12 ± 0.087), and was three orders of magnitude lower than the primary production achieved in mats (Mueller *et al.* 2005).

3.2 Glaciers and ice sheets

Glaciers are often characterized by a significant altitudinal range, so a distinct feature of ecosystem development during the melting period is the gradual transition of the surface from a white snow pack to a darker, debris-rich, ice surface. The same process also takes place on the Greenland Ice Sheet, where marked zones of wet snow and bare glacier ice grow at the expense of the initially dry snow pack (Hodson *et al.* 2010a). The following section therefore begins with a discussion of wet snow as a transient component of the supraglacial ecosystem, before going on to consider the ice surface and the habitats that are exposed by the retreat of the snowline. The role of surface debris known as 'cryoconite' is given particular attention, because it has been the focus of much research and is analogous to the mat-like structures that from on productive ice shelves.

3.2.1 Supraglacial habitats

3.2.1.1 Wet snow

In the early melt period, the initial snow melt percolates into the cold snow cover and refreezes, often forming ice layers and ice columns, and releasing latent heat that warms the pack at depth. Gradually, the snow pack warms, or ripens, until the temperature of the snow is mostly close to the melting point throughout. Further snow melt then soaks the underlying snow, initially ponding above ice layers and then the glacier surface, and so forming slush (Figure 3.5). Flow rates through the dry snow pack are low initially, but speed up as wet snow in flow fingers, which penetrate the snow pack, metamorphose into larger crystals. The coarser wet snow drains more quickly and captures water from the surroundings. Hence, snow melt increasingly flows though these preferential flow paths to the base of the snow pack. Preferential flow paths form here too, at locations controlled by the topography of the ice surface. High nutrient concentrations can be associated with such flows, as found on the Tuva Glacier on Signy Island (Hodson 2006). Snow melt flowing downhill liberates energy as the result of the conversion of potential energy to frictional heat that opposes the flow. Hence, more flow results in more melting out of the flow paths, so the flow paths through and at the base of the snow pack become increasingly larger and more efficient over time.

Wet snow and slush forms a perfect habitat for both algae and phytoflagellates, which in the case of so-called snow algae, can migrate to the snow surface and undergo blooms (see Chapter 2). The biomass of bacteria can be significant, but their activity in glacial snow packs is poorly researched (Amato

Figure 3.5 Snow melt on the surface of Svalbard glaciers during the summer produces relatively thick layers of wet snow or slush in regions of low surface slope. The slush may flow downslope and refreeze on the cold glacier ice beneath the snow pack, forming distinctive mounds, as this image from Polarisbreen shows. From Swisseduc.ch, with permission of Michael Hambrey.

et al. 2007). In some cases, such as on maritime Antarctic glaciers, snow cover persists for much of the melting period and so wet snow is almost certainly the dominant habitat within the entire supraglacial ecosystem. Nitrogen cycling is particularly distinctive in the chemistry of runoff from melting polar glacial snow packs, although it tends to take place at rates far slower than in seasonal snow packs in temperate catchments (Jones 1999; Hodson et al. 2005, 2006). Assimilation of NH_4^+ is particularly apparent, whilst NO_3^- yields tend to be more conservative, although this is not always the case (Fortner et al. 2005; Roberts et al. 2010). NH_4^+ assimilation on glaciers in the maritime Antarctic and Arctic lay in the range of 0.04–0.4 mg NH_4^+ m^{-2} day^{-1}. The greater rate was observed at Tuva Glacier (maritime Antarctic) where emissions from upwind penguin colonies fertilized the snow surface throughout the melt period. The lower rates occurred on the Midtre Lovénbreen and Austre Brøggerbreen in Svalbard (Hodson et al. 2008). These assimilation rates produced estimates of ~0.2–2.3 mg C m^{-2} day^{-1} for gross primary production within the entire supraglacial ecosystem if all other sinks of NH_4^+ within the system were ignored. The upper values, derived for the Tuva Glacier with almost persistent wet snow cover, are about four times lower than the short-term measurements of snow surface photosynthesis derived using ^{14}C-HCO_3^- incorporation rates at almost the same place by Fogg (1967). The impact of microbial processes upon snow pack nutrient economy therefore has significant implications for the magnitude, timing, and composition of nutrient exported to downstream ecosystems, when melting takes place.

The fate of snow pack biomass upon those glaciers where the snowline does retreat up-glacier has not been well studied, but algal blooms can persist upon the glacier ice surface, and discrete cells are also often seen within cryoconite holes. However, detailed comparisons between snow pack and glacier surface microbial communities have been undertaken upon Asian glaciers. For example, Takeuchi et al. (1998) and Yoshimura et al. (1997) established the links between algal biomass and species composition and glacier mass balance in the Himalaya. They found that 'ice environment specialists' (*Cylindrocystis brébissonii*) dominated in a high biomass ablation area, while 'snow specialists'

(*Trochiscia* sp.) dominated algae at higher altitude in the accumulation area. In the intervening part of the glacier, near the equilibrium line, they found that more generalist species (e.g. *Mesotaenium berggrenii*) were most successful, although this was where species diversity was greatest (Yoshimura *et al.* 1997). Similar zonation was found on an Alaskan glacier (Gulkana) by Takeuchi (2001), leading him to propose that algal community structure upon glaciers is intricately linked to the glacier's mass balance status. However, in the Arctic, this link is difficult to establish because a greater proportion of the snow pack biomass can be transported by meltwater from the glacier, without interacting at all with the glacier ice surface. This is due to the formation of a superimposed ice layer at the base of the snow pack. This layer reduces the interaction between the snow and glacier surface habitats and so microorganisms are far more likely to be flushed out of the system by larger supraglacial streams. More studies of nutrient dynamics and biological activity during the transition from snow cover to ice cover are therefore required in these environments.

3.2.1.2 Ice surface habitats upon melting glaciers

Ablating ice surfaces become permeable due to the enlargement of intercrystal voids by solar radiation. This produces a habitat known by glaciologists as the weathering crust within the upper 1.85m of the ice (Irvine-Fynn *et al.* 2011) (Figure 3.6). The weathering crust is known for being less dense than the underlying glacier, unless it is saturated with surface meltwater. In this case the term *rotten ice* is sometimes used to explain the ice surface characteristics. Slow water velocities (between 0.00004 and 0.0007 m^{-1}s^{-1}) are characteristic of the weathering crust, which maximizes the opportunity for nutrient uptake by microorganisms. For example, debris such as cryoconite (see below) is often present through the weathering crust and might enhance its formation due to its absorption of solar radiation just beneath the surface of the glacier (Müller and Keeler 1969). Clearly the weathering crust should be regarded as a major habitat in its own right during summer. However, the energy budget of the surface is not always conducive to its formation. Ice surfaces where the weathering crust is not well developed include maritime environments, where heat advection melts the crust away (Figure 3.6) and the so-called blue ice areas upon cold polar glaciers, where a hard, wind-polished, surface ice is more likely (see below). The surface energy budget of the maritime glaciers means that debris is easily dispersed by meltwater, and therefore occupies streams or is 'stranded' upon the surface. On the surface of Midtre Lovénbreen (Svalbard) such streams revealed bacterial abundances of up to 6.12 × 10^4 mL^{-1} and nanoflagellate abundances of up to 12 × 10^2 mL^{-1} that had the same species composition as cryoconite holes. Ciliates and rotifers were also present in these waters (Säwström *et al.* 2002). However, in larger supraglacial channels, there is usually little opportunity for microbial growth and microorganisms are rapidly advected to downstream environments unless benthic sediments are present or debris is acquired from marginal sources (Scott *et al.* 2011).

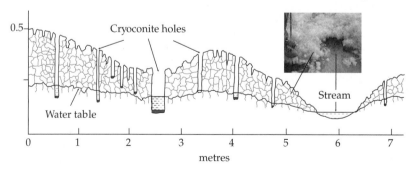

Figure 3.6 Schematic representation of the weathering crust and typical cryoconite holes, as described by Müller and Keeler (1969). The inset shows a typical weathering crust on Midtre Lovénbreen.

3.2.1.3 Near-surface habitats on cold polar glaciers

The hydrology of the cold polar glaciers is confined to the surface, or just below, and follows two modes, dependent on the climate (Fountain *et al.* 2004; Tranter *et al.* 2010). Most of this work has been conducted on glaciers in the McMurdo Dry Valleys. The first mode is characterized by relatively low surface melt and a dominance of subsurface melting as a consequence of the solar heating of debris in the bottom of supraglacial channels, small lakes (cryolakes), and cryoconite holes. Most water drains through a near-surface drainage system that is maintained by the heating of the subsurface debris, and flows into and through cryolakes (see below), which occupy prominent positions in the glacier valley floors (Figure 3.7). The second mode occurs during high melt years. Three have been observed in the last 20 years, in 1085, 1990, and 2001. Much melting of the glacier surface occurs during these relatively warm summers and supraglacial drainage occurs in a similar manner to that seen on High Arctic glaciers in Svalbard. The subsurface drainage network is fully exposed and forms the skeleton of a supraglacial drainage system. The margins of the cryolakes also melt open and become exposed to the atmosphere (Fountain *et al.* 2004). Simultaneously, the lids of the cryoconite holes melt out and are flushed of their aqueous contents. The years

Figure 3.7 Interlinked valleys along the margin of Canada Glacier, Antarctica. These valleys are formed by dust melting into the surface of the glacier. Ice refreezes over the dust as it melts into the glacier surface, protecting the dust from the overlying cold air temperatures. Further solar heating of the dust within the ice produces meltwater that flows beneath the ice surface, ultimately resulting in the formation of the interconnected valleys.

following the warm event in 2001 were notable because of increased primary production in the ice-capped terrestrial lakes in the Dry Valleys. It is hypothesized that dissolved organic carbon (DOC) and nutrients were flushed from cryoconite holes and cryolakes to ephemeral streams that fed the lakes (Foreman *et al.* 2004). Hence, the biogeochemistry and connectivity of cryolakes are important to the functioning of downstream ecosystems. A predicted 2°C warming for the period up to 2050 has been predicted in the sector of Antarctica where the Dry Valleys are located (Shindell and Schmidt 2004). Consequently high melt years will become much more common over the coming decades.

3.2.1.4 Larger supraglacial lakes

The topographic depressions on the surface of the Greenland Ice Sheet near the margins often contain expansive, yet shallow, supraglacial lakes (Figure 3.8). The depressions arise from bedrock topography and ice flow heterogeneity. The lake water is usually dilute meltwater with low suspended solids content. However, Box and Ski (2007) report that cryoconite and other debris can be present at the lake bottom, so a benthic habitat coupled to a oligotrophic pelagic habitat is most likely to exist in these lake ecosystems (Hawes *et al.* 2008). To date there have been few microbiological studies of the larger supraglacial lakes. Smaller, debris-rich 'cryolakes' have been studied upon the surface of cold polar glaciers in the McMurdo Dry Valleys (Figure 3.9). The high sediment content, relatively large reservoir volumes, and presence of microorganisms marks cryolakes as biogeochemically reactive hotspots and prominent, important water bodies on the surface of cold polar glaciers. They are similar to cryoconite holes, but are much bigger and more complex in terms of the heterogeneity of flow paths and sediment thickness, having length scales of 1–20 m. Since they are ice-lidded, their climatic sensitivity means that small changes in annual temperature bring about disproportionate changes in ecosystem characteristics (ACERE 2009). Slight increases in average summer air temperatures may change these frigid systems into more temperate Svalbard- or Greenland-type environments, causing a transition from systems that are relatively isolated or closed from the atmosphere into more open and continually flushed elements of a more temperate supraglacial drainage

Figure 3.8 A supraglacial lake on the Greenland Ice Sheet. This lake is relatively shallow, being confined in a topographical depression. It is fed by numerous small supraglacial streams, but is drained by the large stream in the centre left of the image. Image courtesy of Sarah Das, Wood Hole Oceanographic Institute.

Figure 3.9 A supraglacial lake, or cryolake, on the surface of Canada Glacier, Antarctica. The ice surface is usually present throughout the year.

system (Fountain 1996). Cryolakes are host to a range of microorganisms, dominated by Cyanobacteria, including nitrogen-fixing species, which modify the chemical composition of runoff from frigid polar glaciers (Paerl and Priscu 1998). Thus, just like other supraglacial habitats, cryolakes have the capacity to enhance the overall flux of nutrients from the glacier to downstream ecosystems.

3.2.2 Spatial variations in the biota in supraglacial habitats

There is little information on the variation in diversity or abundances of organisms along supraglacial transects. For example, bacterial abundances on Midtre Lovénbreen showed no significant differences along a transect from an altitude of 200m to the glacier snout, but were significantly lower than those in proglacial lakes close to the snout (Mindl *et al.* 2007) (Figure 3.10). A transect along the Qiyi Glacier in China from 5000 m down to the terminus at 4300 m revealed little difference in the biomass of Cyanobacteria, except where samples were taken on snow rather than ice. Snow samples had the lowest biomass. There was also little difference in the species composition except between samples collected on ice and those collected on snow-covered areas (Segawa and Takeuchi 2010). On glaciers in the McMurdo Taylor Valley there were some biotic gradients depending on elevation and east–west orientation. For example, Cyanobacteria were often present only at the lower and western sides of glaciers, especially so on the Commonwealth, Canada, and Howard Glaciers. Tardigrades were more numerous at lower elevations, while rotifers showed no differences. There were no major chemical differences to explain the observed variations (Porazinska *et al.* 2004). The reasons for these changes most likely indicate that differences between snow cover and bare glacier ice are the greatest source of variability in the microbial consortia, across a melting glacier at a given point in the summer. However, below the snow line in particular, microorganisms upon melting glaciers are well mixed. Further, glaciers are typically surrounded by debris sources that are fine enough to be deposited across the entire ice surface by aeolian processes (usually bringing microorganisms with them). Therefore, larger ice sheets deserve more attention because their scale allows distance–decay studies along transects with increasing distance from the ice margins. The few studies that have been undertaken on the Greenland Ice Sheet reveal clear gradients in the composition and abundance of biologically active surface debris (Hodson *et al.* 2010a; Stibal *et al.* 2010). Along a 50 km transect, Stibal *et al.* (2010) found an increase in the organic carbon and nitrogen content of the debris with increasing distance inland. Maximum values were reached 36 km inland, although levels remained high at 50 km. However, the maximum number of microbial cells and the greatest carbohydrate content of

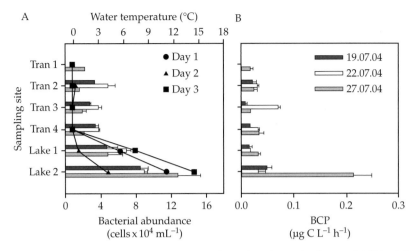

Figure 3.10 (A) Bacteria abundances along a transect down the Midtre Lovénbreen Glacier (tran 1–4) and two proglacial lakes (Lakes 1 and 2) on three successive days shown as columns. (B) bacterial carbon production (BCP). Superimposed are the mean temperature data for the three days. Redrawn from Mindl *et al.* (2007).

the organic carbon were both found at just 6 km inland (Figure 3.11). It is suggested that this zone represents high microbial growth and production, together with ideal conditions for the accumulation of organic matter. Elsewhere, restrictions to the aeolian deposition of organic matter (further inland) and significant removal processes by meltwater flow (nearer the margin) were thought to reduce the biomass and organic carbon storage upon the ice sheet.

Differences between glaciers within the same location do occur and this has been particularly well documented for the Taylor Valley (Porazinska *et al.* 2004). The glaciers are key water sources for life in the soils, lakes, and streams of the Dry Valleys because levels of precipitation are extremely low in what is effectively a polar desert. Five glaciers flow into and terminate in the Taylor Valley; these are the Taylor, Hughes, Howard, Canada, and Commonwealth Glaciers. Invertebrates (tardigrades and rotifers) in cryoconite holes showed considerable variation between glaciers (Figure 3.12). The Hughes Glacier supported no invertebrates and there were significant differences between the other four glaciers. The Hughes Glacier supported a very low diversity of Cyanobacteria compared with neighbouring glaciers and also had no ciliated protozoans. The most biologically active cryoconite holes were found on glaciers on the east side of the Taylor Valley. The results reflected large-scale pro-

ductivity gradients observed in soils and streams in the valley (Alger *et al.* 1997; Virginia and Wall 1999).

One curiosity is that of the so-called ice worms that apparently only occur in the surface ice of coastal glaciers in western North Amercia. They are members of the Phylum Annelida, the segmented worms that also include the earthworms. They are small, up to 15 mm in length, and highly pigmented. They are believed to penetrate irregularities in the ice caused by melting at ice crystal boundaries. The worms emerge at night, thereby avoiding solar radiation (Shain *et al.* 2000). Two species occur, *Mesonchytraeus solifugus* and *Mesonchytraeus rainiernsis*, that fall into two distinct geographical clades—a northern clade that includes all the Alaskan populations, and a southern clade that includes the British Columbia, Washington State, and Oregon State populations (Hartzell *et al.* 2005).

3.2.3 Cryoconite

3.2.3.1 *Structure and formation*

While mats are a common feature on ice shelves, they are uncommon on glaciers and ice sheets. Here steeper surface slopes mean that melt mobilizes surface debris, resulting in its heterogeneous distribution across the ice. In the central parts of the glaciers, away from moraines, debris is dominated by

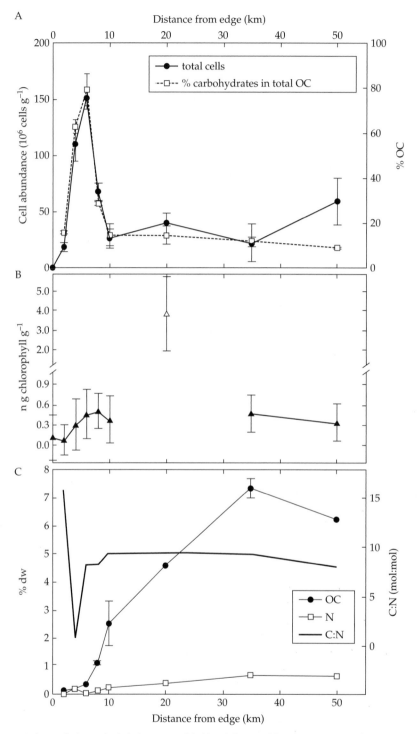

Figure 3.11 Changes in the microbiology and carbohydrate content (A), chlorophyll content (B), and composition (C) of inorganic carbon along a transect across the Greenland Ice Sheet melt zone. The organic carbon content of the debris (OC) is expressed as a percentage of the dry weight (%dw). Redrawn from Stibal *et al.* (2010).

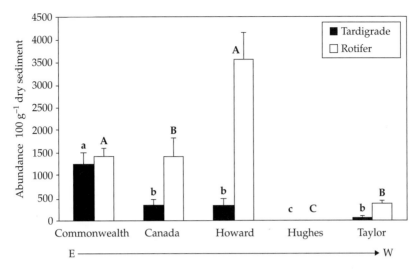

Figure 3.12 Abundances of tardigrades and rotifers in cryoconite holes on glaciers in the Taylor Valley, Antarctica. Capital letters indicate significant differences at $P < 0.05$ among glaciers for rotifers and lowercase letters indicate differences at $P<0.05$ among glaciers for tardigrades. Redrawn from Porazinska *et al.* (2004).

cryoconite, which is often formed by fine, wind-blown particles cemented together into an aggregate particle by microbial filaments and their exudates (Hodson *et al.* 2010b; Takeuchi *et al.* 2010). After ice shelves, the biology of cryoconite is perhaps the most understood component of supraglacial environments. Cryoconite may dispersed across an ice surface, or provide a benthic sediment for streams and lakes. The latter habitats range from the size of cryconite holes (10^{-1} m diameter and depth) to the larger ice sheet lakes described above (10^2 m diameter and 10^1 m depth). The cryoconite hole is by far the most researched habitat on glaciers from both the physical and biological perspectives, however it is only relatively recently that the life cycle of this environment has been described. It begins with the deposition of dust and probably also microorganisms on the snow or ice surface. Some of this material might act as ice nuclei and thus be an integral part of the snow flakes that accumulate to form glaciers and ice sheets (Christner *et al.* 2008). After deposition, melting, perhaps induced by the darkening of the surface by the dust itself, causes a water phase to form around the particle (Bøggild *et al.* 2010). It is at this stage that cell–mineral and mineral–mineral interactions are likely to initiate biofilm formation and eventually form

cryoconite aggregates. For example, hydrophilic and electrostatic forces act rapidly and can bring particles and cells together. Thereafter, the mineral particles become a suitable surface for biofilm formation (Langford *et al.* 2010). Once biofilms have formed, then their continued growth into cryoconite is likely if more sediment and microorganisms are available. The kinetics of this phase are unclear, although robust, spherical aggregates can form very rapidly in ice marginal environments with high sediment supply rates (e.g. in avalanche fans and along the banks of slush-filled supraglacial channels near moraines). At first, biofilms are a distinct feature along the snow line of melting glaciers (Figure 3.13). Further down the glacier, the permeability of the weathering crust of the ice surface provides sufficient time and cell-debris interaction for its transformation into cryoconite aggregates. The growth of the aggregate appears to be driven by primary production, leading to the acquisition of biomass and the production of extracellular polymeric substances (EPS) to glue the particles together (Hodson *et al.* 2010b; Langford *et al.* 2010). At the same time the production of EPS provides an energy substrate for the heterotrophic bacteria. Takeuchi *et al.* (2010) showed particularly well how cryoconite aggregates are effectively 'knitted' together by the high density of near-surface filamentous micro-organisms, usually Cyanobacteria (Figure 3.13). Mobilization (erosion) and growth of these aggregates then results in their spherical, often layered structure (Figure 3.13D). It has been argued that the internal concentric structures in fact represent annual growth layers, suggesting a maximum age of 7 years on the small steep Glacier No 1 in China (Takeuchi *et al.* 2010). However, some of the layer

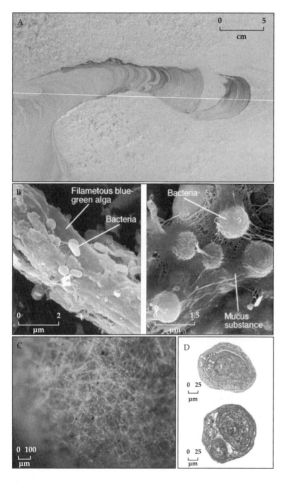

Figure 3.13 (A) Biofilms at the snow line of a melting Svalbard glacier. The surface underneath the snow is a 20 cm thick layer of superimposed ice formed by recent snowmelt refreezing upon the cold glacier surface. Cryoconite aggregate can be found beneath this layer. (B) Left -scanning electron micrograph of bacteria attached to a filamentous cyanobacterium, and right showing mucus-like exopolysaccharide binding. Images courtesy of N. Takeuchi. (C) Densely knitted cyanobacterial filaments upon the surface of cryoconite aggregate from Central Asia. (D) Concentric, internal layers indicate possible annual growth (and/or erosion) stages of cryoconite aggregates. Images C and D from Takeuchi et al. (2010) with permission of the International Glaciology Society.

structures are likely to be destroyed by fluvial erosion thus confounding the dating of grains in this way on glaciers with high melt rates. In many cases the age of the cryoconite is unknown, but it is worth considering that the geochemical conditions within cold, ice-lidded Antarctic cryoconite holes indicate isolation ages of at least ten years (Tranter *et al.* 2004). Furthermore, particle tracking studies on melting glaciers in Svalbard indicate persistence for more than one melt season (Hodson *et al.* 2010b).

Once biofilm becomes cryoconite, there is a major increase in the diversity of the microbial community present. However, it is worth noting that not all cryoconite aggregates are characterized by the robust aggregate spheres described from Central Asia and Svalbard. Observations from Austrian glaciers suggest that cryoconite debris can have a 'smeary' consistency, perhaps due to a higher organic matter content and finer particles (Anesio *et al.* 2010). It remains to be seen whether it is the composition of the debris or the active microbial community that is responsible for this contrast.

3.2.3.2 Biology of cryoconite

Qualitative studies of the organisms inhabiting cryoconite holes date back to the 1930s, but tended to focus on the 'algae' and larger microscopic organisms (Steinbock 1936). The growing interest in the biology of glaciers has now provided more detailed quantitative information on cryoconite communities and their functional dynamics. The best known habitat, a cryoconite hole, resembles a mini-lake with a sediment layer harbouring a benthos and overlying water in which a plankton resides (see Figure 1.21). As is the case in lakes, the sediment of cryoconite holes has much greater diversity than the water column above it (Hodson *et al.* 2008). Those of cold polar glaciers are distinguishable from those upon other glaciers by the presence of an ice lid. The impact of the lid on the exchange of atmospheric gases is important but as yet unquantified with respect to its influence on biological production. However, biological production occurring beneath the ice-lidded cryoconite greatly influences the aqueous biogeochemistry of surface waters on glaciers in the McMurdo Dry Valleys (Tranter *et al.* 2004). The chemistry of open cryoconite holes on Arctic glaciers indicates that they are closer to equilibrium with respect to atmospheric CO_2 than closed Antarctic holes (Hodson *et al.* 2008). Mixing between the plankton and sediment surface communities is also influenced by the presence of an ice lid, because flushing by meltwaters is reduced. However, weak convection overturning the water as debris warms does occur beneath the ice lid (McIntyre 1984).

The communities of cryoconite holes resemble the truncated food webs seen in Antarctic lakes. They are dominated by microorganisms: Bacteria, Cyanobacteria, and phototrophic and heterotrophic nanoflagellates and ciliated Protozoa. Some metazoans such as tardigrades, rotifers, and nematodes live in the sediment of cryoconite holes. Even in the extreme entombed holes of the Antarctic Dry Valleys, rotifers and tardigrades occur, but interestingly no nematodes have been recorded (Porazinska et al. 2004). Nematodes do commonly occur in cryoconite holes in the Arctic and on glaciers at lower latitudes.

The photosynthetic elements present are mainly photosynthetic nanoflagellates, particularly the so-called snow algae (see Section 2.2.1), Cyanobacteria, and diatoms. The latter are not abundant but they are a conspicuous element in the sediment. The culturing of material collected from glaciers in Svalbard and Greenland that had been frozen at –20°C for over a year revealed 27 genera of diatoms. The species were halophitic (salt tolerant), aerophytic (from rock and soil), epipelic surface dwelling, and bryophylic (associated with mosses), suggesting that glacial environments are colonized by diatoms from a wide range of environments, probably by aerial means (Yallop and Anesio 2010). Desmids (Algae) have also been observed in cryoconites on the White Glacier in the High Arctic. The species found were *Ancylonema nordenskiöldii* and *Cylindrocyctis* sp. (Mueller et al. 2001). *Cylindrocyctis* has also been observed in cryoconite holes on glaciers in Svalbard (Stibal et al. 2006). Their presence on Arctic glaciers is not surprising given that desmids occur in Arctic lakes, whereas they are rare in Antarctic lakes and have not been recorded in the Dry Valleys lakes. Consequently desmids have not been reported from Antarctic cryoconite holes.

The diversity of phototrophic nanoflagellates and algae appears high. On a range of glaciers in Svalbard a wide spectrum of species were either observed or isolated from cultured samples (Table 3.4). Many of these species are common in polar lakes. Among the species listed in Table 3.4 are a number of 'snow algae' such as *Chlamydomonas nivalis*. Surface habitats on glaciers can experience high levels of solar radiation.

High photosynthetically active radiation (PAR) can be damaging and inhibit photosynthesis. As indicated in Section 2.2.1, 'snow algae' possess protective pigments that guard against radiation damage.

The concentration of chlorophyll *a*, a measure of photosynthetic biomass, is surprisingly high in cryoconite holes. On the Canada Glacier in Antarctica mean values were $29.1\,\mu g\,L^{-1}\pm20.7$, on the White Glacier in the Canadian High Arctic $124\,\mu g\,L^{-1}\pm95.7$ (Mueller et al. 2001), and on Werenskiolbreen in Svalbard $12–15\,\mu g\,g^{-1}$ of sediment (Stibal et al. 2008). When such concentrations are compared with lake environments in the polar regions the values are high. For example Char Lake and Meretta Lake in the Canadian Arctic had summer chlorophyll *a* concentrations of 0.46–0.78 and $0.96–1.1\,\mu g\,L^{-1}$ respectively, while in Antarctic Lake Fryxell, which is fed by the Canada Glacier, typical values of $1.0–8.0\,\mu g\,L^{-1}$ occurred (Spaulding et al. 1994; Markager et al. 1999). Extreme ultra-oligotrophic Antarctic lakes such as Beaver Lake had chlorophyll *a* concentrations of $0.2–2.7\,\mu g\,L^{-1}$ (Laybourn-Parry et al. 2006). However, the values for cryoconite holes include the sediment, whereas the values given here for lakes refer only to the water column.

The occurrence of Cyanobacteria in cryoconite holes is hardly surprising given their widespread distribution in ice and snow and their extraordinary capacity to successfully colonize extreme environments (see Section 3.2.1). Cryoconite communities are made up of *Phormidium*, *Nostoc*, *Leptolyngbya*, Oscillatoriales, Chroococcales, and Pleurocapsales, depending on location. Svalbard glaciers contained different species of *Leptolyngbya*, *Phormidium*, and *Nostoc* (Stibal et al. 2006), while a glacier in the Qilian Mountains of China was largely colonized by Oscillatoriales with Chroococcales and Pleurocapsales (Segawa and Takeuchi 2010). Molecular analysis showed that 53% of the operational taxonomic units (OTUs) from the Qilian Mountains were similar to those from soil, 24.1% from freshwater, and 2.1% from snow and ice environments, while the remainder were from unknown sources of origin. Seven of the OTUs were similar to those recorded from the Arctic and Antarctic, reinforcing the assertion that the polar regions have Cyanobacteria in common that are not represented at lower latitudes (Jungblut et al. 2010). As in the case of diatoms it appears that glacial habitats are

Table 3.4 Phototrophic nanoflagellates (Protozoa) and Algae either isolated from cultures or observed from cryoconite holes and supraglacial kames on glaciers in Svalbard.

Genus or species	Cryoconite holes	Supraglacial kames
Chlorophyceae spp.	Observed	Observed
Bracteacoccus sp.	Cultured	Cultured
Chlamydomonas nivalis	Observed	Observed
Chlorella homosphaera	Cultured	
Chlorella minutissima	Cultured	
Chlorella vulgaris	Cultured	Cultured
Chlorella cf. chlorococcum	Cultured	
Coleochlamys cuccumis	Cultured	
Klebsormidium flaccidum	Cultured	
Klebsormidium sp.	Cultured	
Muriella terrestris	Cultured	
Muriella sp.	Cultured	
Pseudococcomyxa simplex	Cultured	Cultured
Scotiella sp.		Cultured
Stichococcus cf. chlorelloides	Cultured	
Stichococcus minutus		Cultured
Stichococcus bacillaris	Cultured	Cultured
Tribophyceae sp.	Cultured	
Heterococcus	Cultured	Cultured
Cylindrocystis	Observed and cultured	
Pennate diatoms	Observed	Observed

Data from Stibal et al. (2006).

mainly colonized from terrestrial and freshwater ecosystems.

Heterotrophic bacteria are a major component of the biota of cryoconite holes. Their numbers range widely depending on location and time of year (Table 3.5). As one would expect, abundances are higher in the sediments where there is more organic carbon. The bacterial abundances in cryoconite water and the overlying ice are an order of magnitude lower than reported values from polar lakes. For example in Ossian Sars Lake in Svalbard, across the fjord from the glaciers listed in Table 3.5, bacterioplankton cells reached a maximum of $23.6 \times 10^5 L^{-1}$ and in Lake Fryxell they ranged between 5.3 and $43.6 \times 10^5 L^{-1}$ (Takacs and Priscu 1998; Laybourn-Parry and Marshall 2003). Thus there appear to be some interesting differences between aquatic habitats on glaciers and nearby lakes. While glaciers have high chlorophyll a values and relatively low bacterial abundances, oligotrophic polar lakes have low chlorophyll a but a higher biomass of bacteria. However, given the paucity of data for chlorophyll

a in cryoconite holes, one needs to make comparisons with caution.

Molecular analysis of bacterial communities from cryoconite holes on glaciers in the McMurdo Taylor Valley showed a high percentage of Cytophaga-Flavobacteria in the sediments (up to 87.2%) and a dominance of Betaproteobacteria in the overlying ice (Foreman et al. 2007). Betaproteobacteria also dominate in many snow and ice samples (Alfreider et al. 1996; Battin et al. 2001). On glaciers in Svalbard, Proteobacteria dominated (mostly Alphaproteobacteria), with Bacteriodetes and Actinobacteria contributing 11% and 10%, respectively. Among the Alphaproteobacteria, *Sphingomonas* species were predominant (Edwards et al. 2010). As yet we have limited information on the molecular diversity of heterotrophic bacteria in glacial habitats and indeed polar lakes. However, there is growing interest in exploring this avenue of research as the technology for environmental analysis moves forward, not only in ascertaining what there is, but also in determining what particular groups of bacteria

Table 3.5 Maximum bacterial concentrations in cryoconite holes.

Glacier	Bacteria in sediment	Bacteria in water (x 10^4 mL^{-1})
Canada Glacier, McMurdo DV[1]	20.0×10^4 mL^{-1}	7.9
Hughes Glacier, McMurdo DV[1]	4.5×10^4 mL^{-1}	1.3
Commonwealth Glacier, McMurdo DV[1]	11.5×10^4 mL^{-1}	5.2
Patriot Hills, Antarctica[2]	9.2×10^7 g^{-1}	2.9
Midtre Lovénbreen, Svalbard[3]	7.1×10^4 mL^{-1}	4.5
Midtre Lovénbreen, Svalbard[2]	3.9×10^9 g^{-1}	5.6
Austre Brøggerbreen, Svalbard[2]	9.9×10^8 g^{-1}	7.0
Rotmoosferner, Austrian Alps[2]	3.0×10^9 g^{-1}	10.6
Stubacher Sonnblickkees, Austrian Alps[2]	9.5×10^8 g^{-1}	5.9

Note that in the Dry Valleys (DV) values for water pertain to ice overlying the sediment.
1, Foreman *et al.* (2007); 2, Anesio *et al.* (2011); 3, Säwström *et al.* (2002).

contribute to important processes like biogeochemical cycling.

Heterotrophic flagellated and ciliated protozoans occur in cryoconite holes worldwide. Among heterotrophic nanoflagellates *Paraphysomonas* was common on Midtre Lovénbreen. This is a common genus in lakes and seas worldwide. Concentrations of heterotrophic flagellates ranged between 0.4–4.6×10^2 mL^{-1} in the sediment and 1.1–4.5×10^2 mL^{-1} in the overlying water, but they were always outnumbered by phototrophic nanoflagellates (sediment 17×10^2 mL^{-1} and water maximum 8×10^2 mL^{-1}) (Säwström *et al.* 2002). Ciliated protozoans are observed in cryoconite holes, usually in the sediment, but in most cases no attempt has been made to identify them (Mueller *et al.* 2001; Säwström *et al.* 2002; Porazinska *et al.* 2004). However, Svalbard glacier cryoconites support a small number of species including the haptorid ciliate genus *Monodinium* and the oligotrichs *Halteria* and *Strombidium*, with concentrations in the sediment reaching 10cells mL^{-1} (Säwström *et al.* 2002). These genera occur commonly in the lakes of Svalbard that clearly act as a source of propagules for the colonization of glaciers. *Strombidium* is a ciliate genus that contains species that are well known mixotrophs. In the case of *Strombidium*, mixotrophy involves the sequestration of the plastids from its prey (see Section 1.2.2). It is highly likely that mixotrophic species occur in cryoconite holes.

As indicated in Section 1.2.1 there is now a widespread appreciation of the role that viruses play in aquatic environments in biogeochemical cycling and the transfer of genetic material among bacterial populations. Most of the viruses in the sea and freshwater appear to be bacteriophages, and this is also the case in cryoconite holes. Viral numbers in the water of an Arctic (Svalbard) glacier ranged between 4.86 and 12.71×10^4 mL^{-1}, giving a virus to bacterial ratio of 0.24 to 4.5, which is within the range reported for polar freshwater lakes (Säwström *et al.* 2002; Säwström *et al.* 2008). In common with ultra-oligotrophic freshwater lakes in Antarctica, cryoconite bacteria infected with viruses have low burst sizes. Burst size is effectively an indication of the number of viruses produced by the parasitized bacterial host. On a Svalbard glacier a mean burst size of 3 ± 0.2 viruses per bacterium occurred, which compares with an average burst size of 4 viruses per bacterium for Antarctic freshwater lakes and 26 for lower latitude freshwater systems. Thus in extreme environments where bacterial growth is limited by low temperatures the capacity of an infected host to perform as a 'virus-producing factory' for its parasite is severely curtailed. However, while the production of viruses that can infect new hosts is limited, the percentage of the bacterial community that is infected is much higher in cryoconite holes and Antarctic freshwater lakes and compensates for low burst sizes. The percentage of visibly infected bacteria cells was $11.3 \pm 3.1\%$ and 22.7–34.2%, respectively, while in lower latitude lakes it averaged 2.2% (Säwström *et al.* 2007a). This points to viruses possibly playing an important role in carbon cycling in cryoconite holes, though as yet we have no information on the degree of lysogenic versus lytic cycle (see Chapter 1).

There is a body of evidence referred to above, indicating that supraglacial habitats are colonized by organisms from adjacent terrestrial and freshwater ecosystems. Many of the organisms are capable of forming resting spores or cysts that have evolved to effect dispersal. These are carried by the wind or water. Thus there do not appear to be species endemic to glacier surfaces.

3.2.4 Carbon cycling and biological production

There is debate as to whether supraglacial ecosystems are subsidized ecosystems in terms of carbon cycling or whether they are supported by autochthonous carbon fixation (Laybourn-Parry 2009). As indicated in Section 2.1, Arctic and alpine glaciers in certain parts of the world are likely to receive inputs of carbon and nitrogen from aerial deposition. This material acts as a subsidy to driving biological processes in the same way as occurs in many streams and estuaries, where there are significant allochthonous inputs from a variety of sources. Measurements of the quantity of organic carbon present in the cryoconite of the Wereskioldbreen Glacier in Svalbard was equivalent to 8500–22000 µg C g^{-1} of sediment, while annual primary production was 4.3 µg C g^{-1} (Stibal *et al.* 2008). An accumulation of organic carbon produced during photosynthesis over many years is unlikely, so the obvious implication is that allochthonous carbon derived from the surrounding terrestrial environment by wind transport and aerial deposition from distant sources is contributing to the organic carbon pool. Labile DOC released by autotrophs (see Section 1.2.1) is likely to be low in these extreme aquatic environments. This labile DOC will be taken up rapidly by the bacterial community and will therefore not accumulate. In highly productive environments at lower latitudes, up to 50% of photosynthate may be released into the water during photosynthesis. While we have no estimates for cryoconite holes, this source of DOC is unlikely to be high. The measurement of photosynthesis, bacterial growth, and community respiration are now being conducted in supraglacial ecosystems, but only in a fragmentary nature within cryoconite holes, as the following sections testify.

3.2.4.1 Bacterial production

Bacterial production has been measured on the glaciers of the Taylor Valley, Patriot Hills, and Vestfold Hills in Antarctica, the Arctic (Svalbard), the Greenland Ice Sheet, and on two alpine glaciers in the Austrian Tyrol (Table 3.6). Most of the data refer to the overlying water or ice of cryoconite holes, which as Table 3.5 shows has lower bacterial abundance than the sediment. The limited information from the Dry Valleys indicates that bacterial production is around 95% higher in the sediment than the overlying water, whereas in alpine and Arctic cryoconite holes the difference between the water and sediment was less extreme (Table 3.6). In the Arctic, production in the sediment was between 65% and 81% higher, while on alpine glaciers the sediment was between 35% and 81% higher. However, on the Stubacher Sonnblickkees, minimum levels of production in the water and sediment were similar. Bacterial carbon production in the water of cryoconite holes is exceedingly low compared with other extreme freshwater aquatic environments. The lowest values for both water and sediment in cryoconite holes was recorded during a single incubation survey in late summer on the Greenland Ice Sheet (Cameron 2010) (Table 3.6).

Foreman *et al.* (2007) caution that the rates of production they found on Dry Valleys glaciers may not represent true *in situ* rates. They conducted their experiments in the laboratory under field light and temperature conditions, as did the other investigators quoted in Table 3.6. Bacterial production is measured using ^{14}C-labelled thymidine or leucine or tritiated thymidine. The use of radioactive material in the field is usually not acceptable in Arctic and Antarctic locations for obvious ethical reasons. In the case of the Antarctic studies the overlying ice was melted prior to experiments. Before we can construct meaningful models of carbon cycling for cryoconite holes we will need a much more detailed picture of bacterial growth both in the sediment and water across the entire summer melt phase. The data shown in Table 3.6 were most often derived from assays done on samples taken on one occasion in the melt season.

There are some interesting variations in the size of bacteria found in the water and sediments of glaciers in the Taylor Valley (Foreman *et al.* 2007). There were

Table 3.6 Average ± standard deviation (and range) of bacterial production in the water and sediments of Antarctic and Arctic cryoconite holes.

Location	Production	
	Water	Sediment
	(ng C L^{-1} h^{-1})	(ng C g^{-1} h^{-1})
Antarctic		
Canada Glacier[1]	<50	(350–3329)
Canada, Commonwealth and Taylor Glaciers[2]	0.04 ± 0.02 (0.01–0.06)	23.4 ± 11.8 (9.20–37.1)
Patriot Hills[2]	0.22 ± 0.31 (0.02–0.78)	11.2 ± 4.11 (8.62–19.4)
Vestfold Hills[3]	–	1.58 ± 0.5 (0.97–2.18)
Arctic		
Midtre Lovénbreen, Svalbard[2]	5.27 ± 1.75 (2.88–7.90)	39.7 ± 17.9 (16.9–70.3)
Austre Brøggerbreen, Svalbard[2]	3.26 ± 2.62 (0.50–7.44)	8.62 ± 6.41 (2.06–21.6)
Kangerlussuaq, Greenland Ice Sheet[4]	0.00035 ± 0.00014 (0.00023–0.0058)	0.10 ± 0.030 (0.054–0.14)
Alpine		
Rotmoosferner, Austria[2]	16.7 ± 15.8 (1.20–43.0)	24.6 ± 21.4 (6.00–67.0)
Stubacher Sonnblickkees, Austria[2]	0.05 ± 0.02 (0.03–0.07)	0.13 ± 0.14 (0.02–0.38)

1, Foreman *et al.* (2007); 2, Anesio *et al.* (2011); 3, A. J. Hodson *et al.* unpublished data; 4, Cameron (2010).

significant differences in the size and length to width ratios of bacteria in sediments and water (Table 3.7). The bacteria inhabiting the sediments, where nutrients and carbon substrates are higher, were larger, while those in the ice were very much smaller. Many of the bacteria in the ice may have been in the starvation phase or they may be ultramicrobacteria. Such small bacteria are typically found living within glacial ice (see Section 5.3). Bacteria in the sediment are likely to be associated with or attached to particles. In the water columns of lakes and the sea, many bacteria are associated with particles or aggregates, and these bacteria are larger and have higher rates of production than bacteria freely floating in the water (e.g. Alldredge and Silver 1988; Grossart and Simon 1993; Laybourn-Parry *et al.* 2004). The particles offer enhanced sites of microbial activity and nutrient regeneration. Bacteria from the Canada Glacier cryoconite holes used a range of carbon substrates, of which 29% were amino acids, 35% were carbohydrates, and 24% were carboxylic acids. Photosynthetically produced organic carbon is likely to be the major source of substrate on the Canada Glacier, as in Antarctica aerial carbon deposition is unlikely to be significant.

Table 3.7 Variations in the mean size and length: width ratio of bacteria in the water and sediments of cryoconite holes on glaciers in the Taylor Valley.

	Glacier and location	Cell volume (μm cell^{-1})	Length width ratio
	Canada Glacier: upper west side, ice	0.053	2.37
	Canada Glacier: upper west side, sediment	0.133	5.59
	Canada Glacier: lower west side, ice	0.124	1.90
	Canada Glacier: lower west side, sediment	0.104	1.53
	Hughes Glacier: upper west side, ice	0.149	1.50
	Hughes Glacier: upper west side, sediment	0.549	5.32
	Hughes Glacier: lower west side, ice	0.082	2.17
	Hughes Glacier: lower west side, sediment	0.181	3.38
	Commonwealth Glacier: upper west side, ice	0.051	2.30
SEDIMENT	Commonwealth Glacier: upper west side, sediment	0.129	3.27

Data from Foreman *et al.* (2007).

3.2.4.2 Photosynthesis

Rates of carbon fixation by the photosynthetic communities of cryoconite holes are high compared with polar and alpine lakes (Table 3.8). This includes the water, although by far the highest rates are associated with cryoconite debris. Since cryoconite holes are shallow, PAR penetrates to drive photosynthesis in the sediment. However, in some cases, such as the blue ice areas of the Vestfold Hills (Antarctica), the depth that the debris reaches is as much as 1m, so low light conditions prevail. Figure 3.14 shows PAR penetration through the surface blue ice and also continuous PAR received at the surface of the debris in an 84cm deep cryoconite hole. Significant attenuation of light occurs at depth. Figure 3.15 shows a typical photosynthesis light curve for a cryoconite community exposed to a range of PAR intensities. The notable feature of these data is that photoinhibition occurs at light intensities that are around 9% of the surface irradiation and broadly equivalent to the daytime maximum PAR irradiance observed in the hole (A. J. Hodson *et al.* unpublished data). Thus an autotrophic community adapted to low levels of irradiance appears to exist in these deep, lidded, blue ice cryoconite holes. However, cryoconite need not necessarily reside in melt holes. Much of the cryoconite material on the Midtre Lovénbreen Glacier is dispersed within shallow supraglacial streams or just lying on the ice surface. The kinetics of CO_2 diffusion into the water and towards photosynthesizing cells must therefore differ greatly from the situation in cryoconite holes, although as yet there are no measurements. The irradiance levels experienced by autotrophic communities in surface-dispersed cryoconite are likely to be high, so one would expect communities similar to those seen on snow—snow algae (see Chapter 2).

Thermal equilibration of cryoconite debris within holes evens out the distribution of PAR incident upon the debris surface (Cook *et al.* 2010). This is because solar heating of the debris results in lateral (as well as vertical) heat conduction. When cryoconite material is added to a hole there is a rapid lateral expansion by melting. The cryoconite then disperses over the larger area of the hole and the shading effect of the debris is reduced. Incubations of debris of different thicknesses revealed a shift

Table 3.8 Average ± standard deviation (and range) rates of carbon fixation through photosynthesis in cryoconite holes, and lakes for comparison.

Location	Water	Sediment
	(μg C L^{-1} day^{-1})	(μg C g^{-1} day^{-1})
Antarctic glaciers		
Canada Glacier, Dry Valleys[1]	–	(0.4–1.4)
Vestfold Hills[2]	–	2.10 ± 1.5 (0.21–4.82)
Arctic glaciers		
Midtre Lovénbreen, Svalbard[3]	(0.24–10.56)	(0.63–156.99 μg C L^{-1} day^{-1})
Midtre Lovénbreen, Svalbard[4]	79.8 ± 75.9 (5.38–234)	353
Austre Brøggerbreen, Svalbard[4]	87.5 ± 56.0 (24.8–158)	48
Vestre Brøggerbreen, Svalbard[4]	94.6 ± 58.0 (41.9–190)	208
Longyearbreen, Svalbard[5]	37.5 ± 26.9 (up to 60.0)	17.2 ± 9.7 (2.09–25.7)
Frøya Glacier, Greenland[4]	53.5 ± 59.7 (7.97–183)	115
Kangerlussuaq, Greenland[5]	–	18.7 ± 10.1 (4.7–28.7)
Alpine glaciers		
Stubacher Sonnblickkees, Austria[4]	–	147
Lakes		
Beaver Lake, Antarctica[7] (over summer)	2.1–6.7	
Crooked Lake, Antarctica[8] (over year)	0–38.4	
Char Lake, Arctic[9] (over summer)	3.36–15.6	
Ossian Sarsfjella Lake, Svalbard[10] (over summer)	3.1–7.9	

1, Bagshaw *et al.* (2011); 2, Hodson *et al.* (2010a); 3, Säwström *et al.* (2002); 4, Anesio *et al.* (2009); 5, Hodson *et al.* (2010b); 6, A. J. Hodson *et al.* unpublished data; 7, Laybourn-Parry *et al.* (2006); 8, Henshaw and Laybourn-Parry (2002); 9, Markager *et al.* (1999); 10, Laybourn-Parry and Marshall (2003).

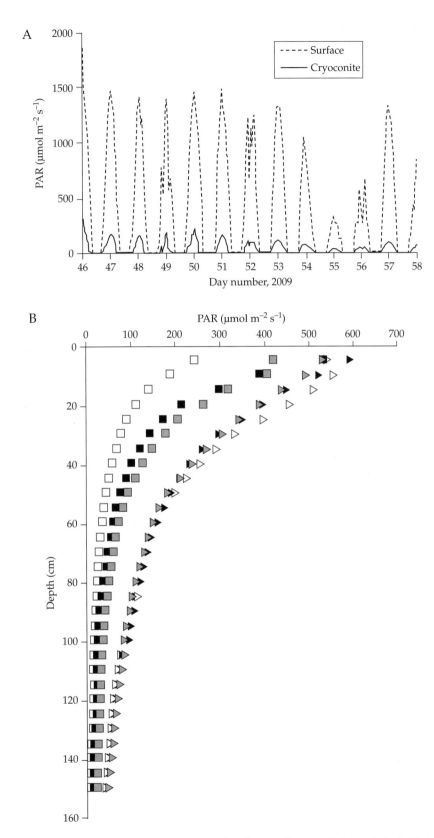

Figure 3.14 (A) Photosynthetically active radiation (PAR) incident at the surface of an 84 cm deep cryoconite hole on the Sørsdal Glacier in the Vestfold Hills, Antarctica. PAR at the ice surface is also shown. (B) PAR profiles with depth in three small valleys and on three ridge crests in the same region. Note how the attenuation and variability of light in the valley profiles (squares) is greater than upon the ridge crest profiles (triangles) as a result of patchy snow cover accumulating in topographic depressions. From A. J. Hodson *et al*. unpublished data.

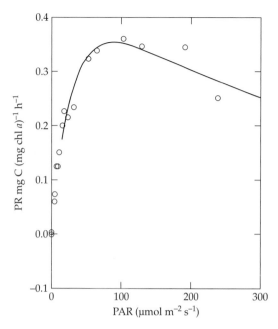

Figure 3.15 A light response curve showing the photosynthetic rate of cryoconite debris sampled from the hole monitored in Figure 3.14A and exposed to a range of irradiances in the laboratory. Note that the maximum rate occurs at irradiance levels broadly similar to the daily maximum levels observed within the hole according to Figure 3.14. PAR, photosynthetically active radiation; PR, photosynthesis rate. From A. J. Hodson et al. unpublished data.

from net heterotrophy (with thick sediment) to net autotrophy (with thin sediment), the change occurring at thicknesses equivalent to 1 or 2 mm. These thicknesses are broadly equivalent to the size of smaller cryoconite grains. Therefore, lateral conduction will attempt to establish a single grain layer across the floor of a stable cryoconite hole and both the debris layer thickness and the net carbon balance will depend upon the size distribution of the cryoconite grains within the hole. The equilibrium both vertically and laterally means that there are far more complex controls on PAR irradiance of photosynthesizing cells in cryoconite holes compared with water bodies like lakes (Gribbon 1979; Cook et al. 2010).

Changes in rates of carbon fixation occur over the melt season. On Midtre Lovénbreen, the highest rates in July occurred in the water, but by early August they had dropped to a low level, while sediment primary production increased progressively over August (Säwström et al. 2002) (Figure 3.16).

There is no obvious explanation for this pattern, as nutrients were not limiting. The other data shown in Table 3.8 are single measurements, often without an indication of when they were made, so there is no basis for establishing seasonal trends elsewhere.

Unfortunately most reported data for photosynthesis do not include measurements of chlorophyll *a* or PAR. Thus it is not possible to ascertain the assimilation numbers or photosynthetic efficiency of the cryoconite autotrophic communities. Assimilation numbers or chlorophyll-specific photosynthetic rates are defined as the units of carbon fixed per unit of chlorophyll *a* (e.g. $\mu g\,C$ fixed $(\mu g\,Chl\,a)^{-1}\,h^{-1}$). The mean value for freshwater plankton is $0.22 \pm 0.22 g C\ (g\ Chl\,a)^{-1}\,h^{-1}$, while for marine plankton it is much higher at 1.38 ± 0.89 (Markager et al. 1999). Polar freshwater lakes appear to have higher values than those of lower latitudes, for example lakes in the Canadian High Arctic had a mean value of $0.45 g C\ (g\ Chl\,a)^{-1}\,h^{-1}$ (Markager et al. 1999) and Antarctic freshwater lakes ranged from 0.015 to $17.3 \mu g C\ (\mu g\ Chl\,a)^{-1}\,h^{-1}$ (Henshaw and Laybourn-Parry 2002). It is likely that cryoconites will reflect the pattern seen in polar lake systems. However, none of the chlorophyll *a* values reported for cryoconites include a correction for degradation products, so active chlorophyll *a* may be lower than the figures suggest.

We also have no figures for photosynthetic efficiency in cryoconite holes. Photosynthetic efficiency is calculated as units of carbon fixed (unit of chlorophyll $a)^{-1}h^{-1}$ (μmol photons $m^{-2}s^{-1}$). Clearly nutrient availability is likely to play a major role in determining photosynthetic efficiency, as well as acclimation to continuous high or low light climates. Again, freshwater phytoplankton peak at the lower end of the spectrum of reported values: $1.23 \pm 0.56 g C\ (g Chl\,a)^{-1}\,h^{-1}\ \mu mol\ m^{-2}s^{-1}$ (range 0.27–5.6) for photosynthetic efficiency in freshwater plankton, and $6.12 \pm 4.8 g C\ (g\ Chl\,a)^{-1}\,h^{-1}\ \mu mol\ m^{-2}s^{-1}$ (range 0.24–35.6) for marine phytoplankton (Markager et al. 1999). In Canadian High Arctic lakes during summer the range was $0.53–2.30 g C\ (g\ Chl\,a)^{-1}\ h^{-1}\ \mu mol\ m^{-2}s^{-1}$, in Antarctic Crooked Lake it was $0.02–5.19 \mu g C\ (\mu g\ Chl\,a)^{-1}h^{-1}\ \mu mol\ m^{-2}s^{-1}$ over the year, and in Beaver Lake during summer it was $0.002–0.39 \mu g C\ (\mu g\ Chl\,a)^{-1}h^{-1}\ \mu mol\ m^{-2}s^{-1}$ (Markager et al. 1999; Henshaw and Laybourn-Parry 2002; Laybourn-Parry

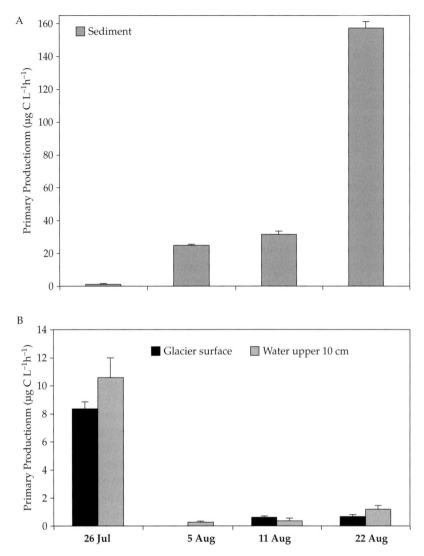

Figure 3.16 Seasonal changes in photosynthetic rate in the sediment at the bottom of cryoconite holes (A) and on the surface of the Midtre Lovénbreen Glacier (Svalbard) and in the top 10 cm of the water in cryoconite holes (B). Data from Säwström *et al.* (2002).

et al. 2006). As with assimilation numbers, one might speculate that cryoconite hole photosynthetic communities are likely to lie within the range reported for polar freshwater systems.

3.2.4.3 Respiration

While bacterial and primary production have been assessed in cryoconite holes, bacterial respiration has not been investigated. There are values for community respiration (Table 3.9), which includes all organisms, but their relative contribution remains unknown. However, a key feature of the available data set is that community respiration rates are broadly similar to primary production. Furthermore, a strong correlation between primary production and community respiration is usually found (Hodson *et al.* 2010a, 2010b; Telling *et al.* 2010). Since the rates of community respiration are much higher than bacterial production (see Table 3.6), it is also unlikely that bacterial respiration contributes much to respired carbon (CO_2). For example, estimates of bacterial respiration have been derived

Table 3.9 Average ± standard deviation (and range) rates of community respiration in cryoconite water and debris, as deduced from dark chamber incubations.

	Water	Sediment
	(µg C L^{-1} day^{-1})	(µg C g^{-1} day^{-1})
Antarctic glaciers		
Vestfold Hills[1]	–	1.86 ± 1.51 (0.40–4.54)
Canada Glacier, McMurdo Dry Valleys[2]	–	2.2 (0.75–2.3)
Arctic glaciers		
Midtre Lovénbreen, Svalbard[1,3]	–	28.2 ± 4.37 (21.9–34.6)
Longyearbreen, Svalbard[1]	3.69 ± 16.9	19.2 ± 5.5 (11.0–25.1)
Austre Brøggerbreen, Svalbard[1,3]	72.9 ± 29.8	15.3 ± 5.02 (6.23–28.6)
Vestre Brøggerbreen, Svalbard[1,3]	–	34.3 ± 2.18 (32.0–37.9)
Kangerlussuaq, Greenland[4]		20.9 ± 8.3 (12.2–29.6)
Alpine glaciers		
Stubacher Sonnblickkees, Austria[1,3]	86.7 ± 17.9	42.1 ± 7.91 (29.7–44.8)

1, Hodson *et al.* (2010b), 2, Bagshaw *et al.* (2011); 3, Anesio *et al.* (2009), 4, Hodson *et al.* (2010a).

using bacterial production and growth efficiencies: BGE = BP/(BR + BP), where BGE is bacterial growth efficiency, BR is bacterial respiration, and BP is bacterial growth or production. A figure of 40% for BGE has been widely used when making such calculations, but its validity across a range of environmental conditions has not been tested. At the extremely low temperatures experienced in cryoconite holes (close to freezing) the cost of maintenance (i.e. BR) is likely to be high so that the energy available to partition into growth will be low. In most freshwater and marine systems, BGE ranges from less than 10% to 25%. However, given the low rates of bacterial production in cryconite holes, even a BGE at the lower end of the range (i.e. 10%) will still mean that bacteria contribute only a small proportion of the overall community respiration rates shown in Table 3.9.

Other sources of respired carbon therefore require examination. There is an apparent imbalance between bacterial respiration and phytoplankton production in unproductive non-glacial aquatic systems, suggesting that these are net heterotrophic systems (del Giorgio *et al.* 1997). Whether or not cryoconite is also a net heterotrophic system therefore remains to be seen. Tables 3.8 and 3.9 indicate a close balance between community respiration and photosynthesis over short intervals (one to three days), but as indicated above we need more data that cover an entire summer season. We also need to identify the contribution of autotrophic and heterotrophic organisms to respired inorganic carbon flux. Another important element in carbon cycling may be the activity of viruses that short circuit the microbial loop, rapidly recycling organic carbon to the pool.

Figure 3.17 (A) Large lateral moraines are a distinct feature of Svalbard's retreating glaciers (photo courtesy of A. Nowak). (B) A supraglacial kame deposit upon Midtre Lovénbreen, Svalbard. The feature was initially a stream deposit, prior to stream migration. The relief then changed because the debris suppressed the ablation of the underlying ice.

3.2.5 Other debris habitats, including the ice margin

Lateral moraines (Figure 3.17A) become wet during the melt season and the fine-grained material is potentially a good substrate for microbial activity. Like any other system, where there is free water life will become established, even as sparse communities. The freshly comminuted and/or exposed surfaces of the debris contain nutrients and reduced compounds that can participate in REDOX reactions, which if microbially catalyzed, provide energy for microorganisms. For example, sulphide oxidation occurs in the wet moraine of Longyearbreen, a valley glacier in Svalbard, which is likely to be microbially mediated (Yde *et al.* 2008). Moreover, the $\delta^{18}O$ signature of NO_3^- and SO_4^{2-} in marginal ice moraines of the maritime Antarctic Tuva Glacier are indicative of bacterial metabolism, especially in the earliest stages of the ablation season (Hodson *et al.* 2009). However, these environments have been shown to support very low levels of biological activity compared with cryoconite holes. For example, bacterial abundances were nearly 70 times lower than in neighbouring cryoconite holes (0.5×10^4 cells mg^{-1}) on the Werenskiold Glacier in Svalbard. Cyanobacteria and other phototrophs were present at concentrations that were 37 times lower than in cryoconite holes (Stibal *et al.* 2006).

Other relevant features include supraglacial kames (Figure 3.17B), which are mounds composed of sand and gravel formed by glaciofluvial deposition. On Svalbard glaciers kames were shown to contain a number of autotrophic organisms (see Table 3.4), but their diversity was significantly lower than in cryoconite holes. Bacterial abundance was almost six times lower than in cryoconites, 6.2×10^4 cells mg^{-1} dry weight of sediment compared with 32.9×10^4 cells mg^{-1} sediment (Stibal *et al.* 2006). The kame deposit shown in Figure 3.17B was distinctly net heterotrophic according to chamber measurements, and significantly enriched the NO_3^- content of the supraglacial meltwaters around the feature (A. J. Hodson unpublished data).

Rates of carbon cycling associated with the above debris have yet to be measured directly, although there are some data from light and dark chamber incubations using marginal debris. Such debris is usually coarser and derived from subglacial sediment sources, brought to the surface by thrusting and ice ablation. Interestingly, both community respiration and photosynthesis of this debris are significantly lower than in cryoconite. For example, a thrust at the margin of the East Antarctic Ice Sheet in the Vestfold Hills produced rates of community respiration and photosynthesis of 0.40 and 1.1 $\mu g C\ g^{-1}$ sediment day^{-1} respectively (A. J. Hodson unpublished data). Similar debris at the margin of the Greenland Ice Sheet in the Kangerlussuaq area produced rates of 0.07 $\mu g C\ g^{-1}$ sediment day^{-1} for community respiration and 0.15 $\mu g C\ g^{-1}\ day^{-1}$ for photosynthesis (Hodson *et al.* 2010a). The values might be lower than expected because the microbial communities within the debris were adapted to a low REDOX sedimentary environment at the glacier bed, such as those described in Section 5.2. It is evident that in the supraglacial environment cryoconite debris supports the highest levels of biological activity.

CHAPTER 4

Sea and lake ice

4.1 Sea ice

4.1.1 Introduction

As outlined in Section 1.4.1, large amounts of water and heat added to or removed from the oceans and atmosphere during the formation and melting of sea ice moderate the climate of the high latitudes (Wang and Overlord 2009; Kumar *et al.* 2010). Sea ice forms when the sea surface freezes, usually at a temperature of –1.8°C because salt inhibits freezing. The polar oceans freeze over in particular during the dark winter months and at least partially melt again during the summer (see Figure 1.8). The severity of freezing and the extent and duration of melting are clearly dependent on the prevailing climate, and how the climate changes on an annual basis (Zhang *et al.* 2010). Confounding factors such as the temperature of the underlying sea, ocean circulation, and the nature of prevailing wind fields obscure simple and direct associations with climate change to be made (Notz 2009; Kwok and Morison 2011). However, a warming global climate, in general, is decreasing both the area and the duration of Arctic sea ice cover (Kwok *et al.* 2009; Markus *et al.* 2009). By contrast, the area of sea ice in Antarctica might be rising very slightly (Turner *et al.* 2009).

Sea ice forms and melts due to a combination of factors, including the age of the ice, air temperature, and solar insolation (Notz 2009). During the winter, the area of the Arctic Ocean covered by sea ice increases, usually reaching a maximum in March. As the seasons progress, the area covered in sea ice decreases, reaching a minimum in September during most years (Tietsche *et al.* 2011). First-year ice that survives the summer thickens in the following winter to form multi-year or old ice. First-year ice melts more easily than older ice for two reasons,

being both thinner and less permeable than older ice. Summer meltwater tends to form deeper ponds on first-year ice surfaces than on older ice, reducing the albedo of the surface and increasing the capture of solar energy, which in turn results in greater solar energy capture (Yackel *et al.* 2007).

There are many different types of sea ice (WMO 2009). Land-fast ice, or fast ice, is sea ice frozen along the coastline or to the shallow ocean floor, so it does not move under the influence of currents and wind. By contrast, drift ice floats on the sea surface of the water and does move under these influences. Large masses of drift ice are termed pack ice. Pack ice is prone to occasional blockage due to flow restrictions. Large areas of pack ice are found in the Arctic Ocean and the Southern Ocean.

The first sea ice to form in calm water is a surface skim of individual crystals that initially are 2–3mm discs floating flat on the surface. Next, long fragile arms (or dendrites) form that easily break off due to wind or wave activity, leaving a mixture of discs and arm fragments. Further turbulence in the water causes these fragments to break up further into smaller crystals to form a suspension of increasing density in the surface water, giving rise to frazil or grease ice. It is suggested that crystals of frazil ice form on suspended particles in the water column and then rise to the surface. Certainly, newly formed frazil ice contains high concentrations of microorganisms that have been harvested from the water column (Garrison *et al.* 1986, 1989). The water columns of aquatic environments, lakes, or the sea, contain microscopic and macroscopic particles of organic and inorganic matter that form foci for microbial activity. Usually, concentrations of bacteria, microalgae, and protozoa are higher on these aggregates than in the open water and they often achieve higher levels of production than organisms floating freely in the water. Frazil ice formation may also result from cold snow falling on the sea surface, initially forming a floating mass of slush. The frazil crystals freeze together when calmer surface conditions return to form a continuous, thin (a few centimetres) sheet of transparent young ice called dark nilas. The dark nilas thickens and turns first grey and finally white, loosing transparency as it thickens. Further thickening of the sea ice now occurs by water molecules freezing onto the base of the nilas by so-called congelation growth, producing first-year ice with a thickness of up to 2 m.

Sea ice formation in rougher water results in several different intermediate sea ice types en route to the formation of first-year ice. Shuga is formed in agitated conditions by the accumulation of slush or grease ice into spongy pieces several inches in size. Turbulence compresses these ice particles into larger plates, typically 1–3 m in diameter, called pancake ice. These collide together giving the pancakes upturned edges. The pancake plates raft over one another or freeze together to form consolidated ice pancake ice, which has a very rough appearance (WMO 2009).

The annual increase and decrease of polar sea ice has been the subject of intensive research during the last decade in particular. Current projections of sea ice loss suggest that the Arctic Ocean will be ice-free by 2080 (Holland et al. 2006), although this date could be as early as 2030 (Wang and Overlord 2009). This will be in stark contrast to the last 700 000 years when at least some sea ice is believed to have remained during the summer months, even during periods when the Arctic was warmer than it is today. The continued loss of Arctic sea ice and enhanced illumination of surface waters should result in increased marine productivity and draw down of atmospheric CO_2 (Arrigo et al. 2010). However, a number of confounding factors may limit ocean productivity (Bluhm and Gradinger 2008) including a paucity of nutrients and the input of relatively inert dissolved organic carbon (DOC) in runoff from terrestrial environments, which reduces light penetration through surface waters (Arrigo et al. 2010). The Arctic Ocean has freshened over the last decade, with the contribution of terrestrial ice melt to this freshening (an anomaly of

$118 \pm 51 km^3$ year^{-1} compared with earlier decades) approximately equal to the contribution of terrestrial riverine runoff (an anomaly of $104 \pm 50 km^3$ year^{-1}) (Peterson et al. 2006). Current modelling suggests that glacial discharge will increase in the warming Arctic climate over at least the next 50 years (Dyurgerov et al. 2010; Mernild et al. 2010). Glacial runoff is also a supplier of potentially bioavailable nitrogen and phosphorus to the oceans (Hodson et al. 2004; Hood and Scott 2008), and a major source of a very different type of DOC to the refractive forms found in riverine runoff from terrestrial sources. Glacial DOC is labile and is readily available for heterotrophic bacterial production in high latitude coastal waters (Anesio et al. 2009; Hood et al. 2009; Fellman et al. 2010).

The Arctic sea ice September minimum extent reached new record lows in 2002 (Serreze et al. 2003; Stroeve et al. 2007) and 2007 (39.2% below the 1979–2000 average). In 2007 (Kwok et al. 2009), Arctic sea ice broke all previous records by early August—a month before the end of melt season—with the biggest decline ever in the Arctic sea ice minimum extent of more than a million square kilometres. The legendary Northwest Passage opened completely for the first time in human memory and was widely reported in the media. In 2008 and 2009, the area of Arctic sea ice at the minimum extent was higher than in 2007, but it did not return to the levels of previous years. The remaining ice is both thinner and younger than in previous years (Tietsche et al. 2011).

The rate of the decline in Arctic sea ice area is accelerating. This was ~3% per decade during 1979–1996, compared with ~11% per decade for 1999–2008. There was a huge loss of multi-year ice during 2005–2008, which decreased by ~42% in area and ~40% in volume (equivalent to ~6300 km^3) (Serreze 2011; Stroeve et al. 2011).

Sea ice around Antarctic shows the same seasonal pattern of growth and retreat as in the Arctic, but in marked contrast to the annual declines shown in the Arctic, the area of sea ice around Antarctica has grown by a small, but statistically significant, amount since the late 1970s (Turner et al. 2009). Antarctica as a whole is believed to have warmed slightly over the last two decades, so this is a surprising result. The largest increase in the area of sea ice occurs in the autumn, and this occurs most significantly in the Ross Sea. By contrast, there is a marked loss of sea ice area in the Bellinghausen Sea. The net effect is, however, one of an increased area of sea ice. The reason for the increased area of sea ice could be due to the depletion of stratospheric ozone, which affects the atmospheric energy balance and the location of the regional wind fields (Turner *et al.* 2009). A significant impact is that there are increased wind speeds in the autumn over a significant region of the Southern Ocean, which introduces colder air to these areas, so resulting in increased sea ice formation (Schlosser *et al.* 2011). However, this effect is of the same order as natural climate variability in the region, so the ozone depletion effect should be regarded with caution (Turner *et al.* 2009).

As indicated in Section 1.4.1, sea ice is a complex structure with a matrix made up of numerous channels filled by brine. However, there are differences between land-fast ice, first-year and multi-year ice, pack ice, and the marginal ice zone (the region where sea ice meets open water). The distribution of organisms and their abundances reflect the different -environmental conditions that prevail in these different forms of sea ice. Pack ice is that ice that floats freely on the sea and is not attached to the land. In pack ice assemblages of organisms occur in the surface layer and in internal pockets, while in

land-fast ice the organism mainly occur in a bottom layer close to the water–ice interface. In the pack ice there is little seasonal change in chlorophyll *a* concentrations in the internal pockets, while in the surface layer there are marked seasonal differences with high peaks (exceeding 400 µg chlorophyll *a* L⁻¹) in late summer. The surface layer is the major contributor to primary production in the pack ice (Garrison and Buck 1991) (Table 4.1). The growth of the sea ice algae in the various pack ice habitats depends not only on the light climate but also on the availability of nutrients that are essential for photosynthesis. Surface flooding of the ice plays an important role in providing nutrients to the surface communities. This process involves snow loading of the ice and submersion of the pack ice. It is estimated that surface flooding occurs over 15–30% of the pack ice. Thus while snow cover strongly attenuates light it can have a positive effect in facilitating flooding and replenishment of nutrients from seawater. Figure 4.1 illustrates the tight coupling between primary production and snow depth in Weddell Sea pack ice (Arrigo *et al.* 1998). During the peak of primary production in January and February decreases in snow thickness elicited a sharp drop in production. This reduction was due to nutrient limitation. The replenishment of nutrients can only occur when the snow cover is thick enough to produce surface flooding. Thick snow cover weighs

Table 4.1 Rates of primary production and chlorophyll *a* concentrations in sea ice from the Arctic and Antarctic.

Location	Chlorophyll *a* (mg m⁻²)	Primary production (mg C m⁻² day⁻¹)	Source
Antartic			
Prydz Bay pack ice (S)	13.7	63.4	McMinn & Hegseth 2003
Weddell/Scotia Sea pack ice (Su)	0.5–23.6 µg L⁻¹	0.7–113.6	Garrison & Buck 1991
Weddell Sea pack ice (A)	260 µg L⁻¹ max	6–300	Lizotte & Sullivan 1992
Weddell Sea multi-year ice (A/early W)	0.7–11 mg m⁻³	22–176	Fritsen *et al.* 1994
Eastern pack ice (Su)	2.5–3.4	13.4 max	McMinn *et al.* 2007
West of Antarctic Peninsula fast ice (W/early S)	2.0–11.6	24.7–60	Kottmeier & Sullivan 1987
McMurdo Sound fast ice brine (Su)	0.7–1.6 µg L⁻¹	0.4–12	Stoecker *et al.* 2000
Prydz Bay fast ice (Su)	14.9–36.9	94–637 m⁻³ day⁻¹	Archer *et al.* 1996
Arctic			
Svalbard Bank pack ice (S)	0.19	12.0	McMinn & Hegseth 2003
Barents Sea pack ice (late S)	6.9–19.5	9.6–67.2	McMinn & Hegseth 2007
Off Labrador pack ice (W)	40–191 mg m⁻³	16.8–187.2	Irwin 1990

S, spring; Su, summer; A, autumn; W, winter.

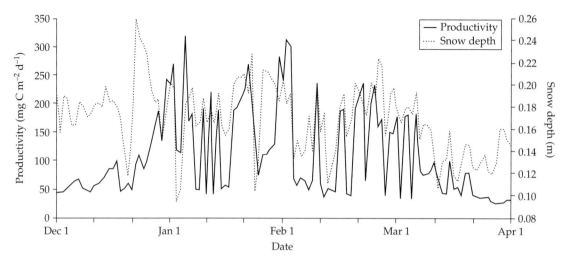

Figure 4.1 Modelled coupling between primary production and snow depth during the phase of maximum algal growth between January and February in the Weddell Sea. Redrawn from Arrigo *et al.* (1998).

down the ice so that it floats lower in the water, permitting flooding by seawater.

Land-fast ice communities dominate the bottom of the ice column close to the sea–water interface. While the light climate is poor, nutrients are continually supplied by seawater that penetrates the brine channels. As indicated in Section 1.4.2, the sea ice algae are extremely well adapted to low levels of photosynthetically active radiation (PAR), and provided nutrients are available are able to sustain photosynthesis. The ice edge margin where the ice meets open sea is a region of high productivity in the open water, which is influenced by the ice. The phytoplankton in this region develop in a stable, surface layer of water of low salinity from ice melt. The Southern Ocean has a deep mixed layer and seldom has any stratification because frequent storms prevent its formation. The low salinity, low density meltwater provides a stable layer enabling the phytoplankton to grow in a well-illuminated environment. Such ice edge blooms are very important in the overall productivity of the Southern Ocean. Nutrients are usually not limiting in this situation; however, in the Arctic, upwelling at the ice edge may be an important mechanism for meeting nutrient demand. The ice edge also serves to reduce wind-driven turbulence. Further, the development of blooms may be enhanced by the seeding

of the phytoplankton from the ice communities (Smith and Nelson 1986).

Multi-year ice is ice that covers the sea continuously over years. As indicated above, there are growing concerns about the loss of this type of ice cover, particularly in the Arctic. Multi-year ice occurs in parts of Antarctica, such as the western sector of the Weddell Sea. Where multi-year ice occurs there is little productivity in the underlying water because little light penetrates the ice to drive photosynthesis. A study in this part of the Weddell Sea, which deployed a drifting ice station, provided a very good picture of the biological activity in the ice column (see Section 4.1.3). In this ice the major zone for algal growth was in the upper 0.4 m of the ice near the porous region (Fritsen *et al.* 1994, 1998) (Figure 4.2). As the figure shows, the ice profile includes 5–10 cm of consolidated granular ice above 10 cm of unconsolidated ice crystals and seawater. The upper 5–10 cm of granular ice is thought to have arisen through the freezing of a seawater–snow mixture. It is estimated that during the initial period of algal growth, fluxes of seawater needed to exceed $0.015 \text{m}^{-3} \text{ m}^{-2} \text{ day}^{-1}$ in order to supply the nutrients needed to sustain production. The hypothesis is that the porous layer (Figure 4.2) must have had its original brine replaced up to 20 times during the autumnal freezing period, probably through the

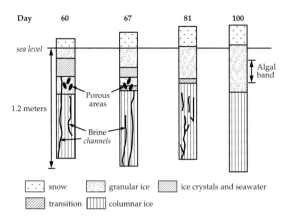

Figure 4.2 Diagrammatic representation of the structure of multi-year ice from the Weddell Sea between March and June. Redrawn from Fritsen *et al.* (1998).

vertical brine channels. Additional nutrients may be provided by flooding caused by surface snow, as outlined above for pack ice (Fritsen *et al.* 1994).

4.1.2 Adaptations

As indicated in Section 1.4.2, the sea ice is a highly productive environment despite having a poor light climate. It is the domain of extremophiles and presents organisms with a range of physiological challenges, among which constant low temperatures, variations in salinity, and low levels of PAR are the most important.

PAR is extremely low in the ice; as little as 0.1% of surface irradiation reaches the diatom communities that live in the bottom ice layers of land-fast ice. Light levels of around 15 μmol photons m^{-1}s^{-1} are typical. Photosynthesis has even been detected at irradiances as low as 1 μmolm^{-1}s^{-1} (Palmisano and Sullivan 1983). Sea ice algae are arguably among the most shade-adapted plants on Earth. The shade adaptation indices (E_k) are less than 15 μmol photons m^{-2}s^{-1} and there is evidence of photoinhibition as low as 20 μmol photons m^{-2}s^{-1}. In the short term, adaptation is mediated by a downregulation of photosynthesis when PAR is too high, and in the longer term by changes in the photopigment concentrations, for example fucoxanthin which is very effective at absorbing the wavelengths that penetrate ice. During April and May in an Arctic sea ice community both chlorophyll *a* and carotenoids

decreased while the ratio of carotenoids to chlorophyll *a* increased over a period when irradiance was increasing (Michel *et al.* 1988). Over a diurnal cycle, sea ice communities rapidly adapted to increasing or decreasing levels of irradiance (McMinn *et al.* 2003) (Figure 4.3). The study from which the data in Figure 4.3 were derived, suggested that the algae never reached light saturation at ambient irradiances. In other words they were continually light limited.

Sea ice algae subjected to differing light intensity and low temperatures under laboratory conditions appear to develop psychrophilic acclimation. The algae have the capacity to photosynthesize at freezing temperatures even when light intensity increases. In the natural environment where seasonal light changes are gradual, this allows for progressive selection of species that are more adapted. Psychrophilic adaptation is probably mediated by increases in photosynthetic enzymes with decreasing temperature (Rochet *et al.* 1985). Investigations of such enzymes in psychrophilic *Chloromonas* species suggest a possible novel adaptation to the cold, where increased production of key enzymes like RUBISCO (ribulose-1,5-bisphosphate carboxylase/oxygenase) counterbalances poor catalytic efficiency at low temperatures (Devos *et al.* 1998). RUBISCO is an enzyme involved in the Calvin–Benson cycle or the photosynthetic carbon reduction cycle. It catalyses the first major step in the process of carbon fixation. It can constitute 50% of the soluble protein in algal cells and has been said

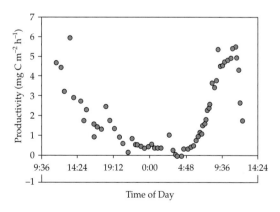

Figure 4.3 Diurnal pattern of primary production in sea ice off eastern Antarctica over a 30 h period. Redrawn from McMinn *et al.* (2007).

to be the most abundant enzyme in the world. RUBISCO also catalyses a carboxylase reaction and an oxygenase reaction (Falkowski and Raven 1997). In their comparative study of psychrophilic and mesophilic algae, Devos *et al.* (1998) found that the psychrophilic algae contained significantly higher amounts of RUBISCO than their mesophilic counterparts. In many cases cold-adapted enzymes exhibit modifications of their kinetic parameters, allowing them to catalyse reactions at low temperatures. In the case of RUBISCO this did not appear to be the case, instead the algae produced more of the enzyme to compensate for reduced activity at lower temperatures.

All organisms in the sea ice face continuous low temperatures, which because of the brine are considerably lower than those in seawater. One of the major challenges at low temperatures is the maintenance of the fluidity of lipid membranes. The fatty acid composition of the phospholipids forming membranes is fundamental to regulating the degree of fluidity of the membrane. At low temperatures there is an increase in the proportion of unsaturated fatty acids, a decrease in the average chain length, and an increase in polyunsaturated fatty acids (PUFAs). To date the majority of data on such changes have been derived from studies of bacteria rather than eukaryotes. The synthesis of PUFAs involves pathways catalysed by polyketide synthases (PKSs), and there is evidence to suggest that in cold aquatic ecosystems PKSs have evolved that are distinct in both their structure and mechanism (Metz *et al.* 2001). The capacity of sea ice algae, and probably also bacteria, to produce PUFAs depends on their nutritional status, which is determined by a range of factors among which light (for autotrophs) and nutrient availability are fundamental. Sea ice algae concentrations of PUFAs were negatively correlated with irradiance levels, so that algal PUFA concentration declined on average by 40% between April and June in the High Arctic (Svalbard) (Leu *et al.* 2010). This also has importance consequences for the organisms that are nutritionally dependent on the sea ice algae, as reduced PUFA content renders the algae of lower nutritional value.

Variations in salinity occur within the sea ice. As ice forms salts are excluded and the salinity of the surrounding water increases, in some cases up to three times that of seawater. Equally when the ice starts to melt there is a flux of freshwater that reduces salinity below that of seawater. Such variations impose significant osmotic stress on the organisms inhabiting the sea ice. Many ciliated protozoans can deal with the challenge of hyposaline conditions by virtue of their osmoregulatory capacity. They possess a structure called the contractile vacuole, surrounded by a system of fine membranous tubules and vesicles termed the spongiome. Through this system, water that enters the cell through the cell membrane by osmosis is concentrated into the contractile vacuole, which periodically contracts, expelling the collected water from the cell. Marine flagellates and amoebae apparently lack contractile vacuoles (Laybourn-Parry 1984). In bacteria lipid packing and the proportion of fatty acids are regulated and a range of halotolerant enzymes able to function over a range of salinities have been identified. In some sea ice, algae produce osmolytes (e.g. mannitol, proline) under hypesaline conditions, which are broken down under hyposaline shock. Among these osmolytes is dimethylsulfoniopropionate (DMSP), which is also a cryoprotectant. This is found in sea ice algae and is the precursor to dimethyl sulphide (DMS), which is released on cleavage of DSMP into the atmosphere where it pays an important role in cloud formation (Thomas and Dieckmann 2002). However, information on how many organisms deal with the challenge of varying salinity in cold environments is fragmentary.

4.1.3 Community structure and production

The sea ice community is one of the most productive and structurally complex seen in ice and snow environments (Figure 4.4) (see Section 1.4.2). It includes a range of metazoans such as krill, copepods, rotifers, and turbellarians, and through important species like krill links into the main marine food web. The autotrophic community is usually dominated by diatoms, but autotrophic nanoflagellates and dinoflagellates may contribute significantly to photosynthesis. The heterotrophic protozoan community includes nanoflagellates, dinoflagellates, heliozoans, and ciliates.

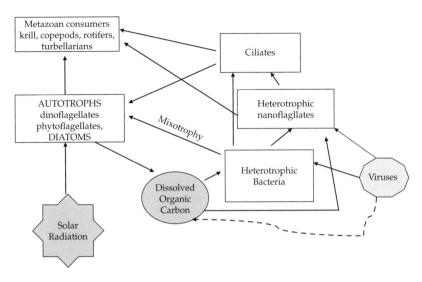

Figure 4.4 Diagrammatic representation of the sea ice community.

4.1.3.1 Primary production

Diatoms dominate the autotrophic communities of the sea ice and can reach levels of 1000 µg chlorophyll a L^{-1} (see Figure 1.4). As indicated above, they are mostly located in the lower few centimetres of the land-fast sea ice and in the upper layers of ice floes and multi-year ice. Levels of chlorophyll a in the Southern Ocean water column range between 0 and 5 µg L^{-1}. In broad terms the Antarctic sea ice achieves an annual level of primary production between 63 and 70 TgC, which is estimated to contribute around 5% of total primary production in the Southern Ocean (1300 TgC year^{-1}). This is a relatively small percentage, but its significance lies in the fact that it occurs at a different time to the peak of primary production in the ocean waters. Sea ice primary production provides an important source of energy to a variety of organisms and in particular krill (*Euphausia superba*)—one of the mainstays of the Southern Ocean food web. The standing stock of krill can exceed 1.5 billion tonnes. They and other microcrustacea are an important food source to squid, penguins, seals, and some whale species (Thomas and Dieckmann 2002). Long-term decreases in the extent of sea ice in the northern and southern polar regions will impact significantly on the availability of winter food for important species like krill and the copepods.

The species diversity of diatoms found in the sea ice is quite high, considering the extreme nature of the environment, but a small number of species tend to predominate. The pack ice communities of the Arctic and Antarctic are remarkably similar in terms of biomass, productivity, and biodiversity at the generic rather than the specific level. For example, in Prydz Bay and the Svalbard Bank, the communities were dominated by *Fragilariopsis curta*, *Fragilariopsis cylindrus*, *Navicula*, and *Nitzschia* in the former and *Nitzschia frigida*, *Fragilariopsis oceanica*, and *Navicula pelagica* in the latter (McMinn and Hegseth 2003). The dominance of particular species holds true across locations in the Southern Ocean. In the Weddell Sea pack ice *F. cylindrus* and *F. curta* contributed the highest proportion of the community during April to December. *F. cylindrus* contributed between 13% and 83% of diatom abundance and *F. curta* between 10% and 11% (Gleitz *et al.* 1998). These data were collected on a cruise 60°S to 75°S and it is worth noting that the majority of information on sea ice assemblages is derived from studies conducted in nearshore environments from research stations on fast and pack ice. In the southern Weddell Sea two distinct diatom assemblages were apparent that differed from one another and from those seen in the outer pack ice zone. In the southern Weddell Sea one set of ice cores was dominated by *F. cylindrus* (mean relative abundance 41%), *Tropidoneis glacialis* (17%), and *F. curta* (11%) while the other set of cores derived further south were dominated by *Thalassiosira antarctica* (44%) with *Chaetocerus* sp., *Porosira glacialis*, and *F. cylindrus* as the major co-occurring species (Gleitz *et al.* 1998). This data set emphasizes larger scale variations that are undoubtedly related to conditions in the sea ice.

However, most studies are undertaken over specific short periods, usually in summer or spring. What the data show are snapshots of part of an annual cycle of sea ice community production and tell us nothing about seasonal successions. Table 4.2 provides details of species found in the sea ice, but should be viewed bearing in mind that most of the data are snapshots.

Table 4.2 Diatoms species reported from sea ice communities in the Antarctic and Arctic.

Species	Antarctic	Arctic
Actinocyclus actinochilus	4	
Amphiprora sp.		5
Amphiprora kjellmanii	4	
Amphiprora kufferathii	2	
Berkeleya rutilans	2	
Chaetocerus spp.	1, 2, 4	5
Chaetocerus dictaeta	2, 4	
Chaetocerus neogracile	2, 4	
Corethron criophilum	1, 4	
Coscinodiscus ocula iridis	1	
Cylindrotheca cloisterium	1, 2	1, 3, 5
Dactylisolen antarticum	1	
Dactylisolen tenuijunctum	2, 4	
Entomeneis sp.		1
Entomeneis kjellmannii	1	
Eucampia antarctica	1	
Fragilariopsis curta	1, 2	
Fragilariopsis cylindrus	1, 2	3
Fragilariopsis kergulensis	1, 2	
Fragilariopsis oceanica		1, 5
Fragilariopsis vanheurckii	2	
Gyrosigma tenuirostrum		1
Navicula sp.		3, 5
Navicula cryophila	2	
Navicula kariana		1
Navicula pelagica		1, 3, 5
Nitzschia spp.	1, 4	3,
Nitzschia arctica		3
Nitzschia closterium	4	
Nitzschia cylindrus	4	
Nitzschia frigida		1, 3
Nitzschia lecointei	4	
Nitzschia neglecta	2	
Nitzschia prolongatoides	4	
Nitszchia promare		1, 3
Nitzschia stellata	1	
Nitzschia subcurvata	1, 4	
Nitzschia turgiduloides	1, 4	

Species	Antarctic	Arctic
Pleurosigma sp.	1	1
Porosira glacialis	2	
Probiscia truncata	1	
Pseudogomphonema artica		1
Pseudo-nitzschia sp.		1, 5
Pseudo-nitzschia prolongatoides	2	
Pseudo-nitzschia subcurvata	2	5
Pseudo-nitzschia turgiduloides	2	
Thalassiosira sp.	4	1, 5
Thalassiosira antarctica	1, 2	
Thalassiothrix spp.	1	
Tropidoneis glacialis	2, 4	

1, McMinn and Hegseth (2003) from Prydz Bay and Svalbard Bank; 2, Gleitz *et al.* (1998) from the Weddell Sea; 3, Rózanska *et al.* (2009) from the Beaufort Sea; 4, Michel *et al.* (1988) from Hudson Bay dominant species only; 4, Garrison and Close (1993) from Weddell Sea pack ice; 5, Werner *et al.* (2007) from pack ice around Svalbard.

The highly productive marginal ice zone phytoplankton blooms are probably seeded by populations of diatoms from the sea ice. *F. cylindrus*, a dominant species in the sea ice community, was found as a major species in the ice edge phytoplankton of the Bellinghausen Sea, the Weddell Sea, and Prydz Bay in spring and summer. Across these locations this species contributed up to 39% (range 6.3–39.3%) of total phytoplankton diatom numbers. *Cylindrotheca closterium*, another sea ice species, contributed up to 44% (range 0.8–43.6%) (Kang and Fryxell 1992).

While most investigations have shown that the sea ice autotrophic community is dominated by diatoms, relatively few studies have considered the entire assemblages of phototrophs. As well as diatoms, autotrophic dinoflagellates are very common, as are phototrophic nanoflagellates. On occasion these groups can dominate over diatoms, as occurred in the Weddell Sea pack ice during November in one study. Here dinoflagellates and the prymnesiophyte *Phaeocystis* dominated. The autotrophic (mixotrophic) ciliate *Mesodinium rubrum* was also present but did not contribute significantly to overall biomass (Garrison and Buck 1991) (Figure 4.5). On the few occasions that the sea ice community has been studied over an annual cycle dinoflagellates have shown a clear seasonal pattern that is related to the availability of PAR. In land-fast ice in Prydz Bay (eastern Antarctica), dinoflagellate numbers dropped to barely detectable in June (when most had encysted)

and appeared again in September when their numbers started to increased substantially (Figure 4.6). In November the small dinoflagellate *Polarella glacialis* was the dominant species. During the winter the number of autotrophic nanoflagellates also decreased, but did not entirely disappear. Among the species were a number such as cryptophytes and *Pyramimonas* that are capable of mixotrophy (Paterson and Laybourn-Parry 2011). These species may have persisted through winter, when photosynthesis was

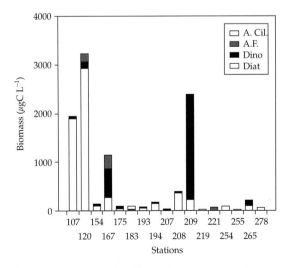

Figure 4.5 Composition of the biomass of autotrophic community at a series of stations in the Weddell/Scotia Sea. A.F., autrotrophic flagellates; A. Cil, ciliate *Mesodinium rubrum*; Dino, dinoflagellates; Diat, diatoms. Redrawn from Garrison and Buck (1991).

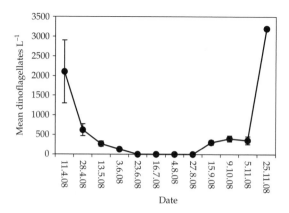

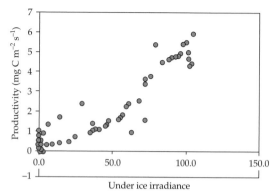

Figure 4.6 Dinoflagellate abundances over an annual cycle in the sea ice in Prydz Bay (Paterson and Laybourn-Parry unpublished data).

Figure 4.7 Primary production in relation to under-ice irradiance in the sea ice off eastern Antarctica. Redrawn from McMinn *et al.* (2007).

not possible, by feeding on bacteria or by taking up DOC. Mixotrophy has been shown to be an important trophic strategy for phototrophic nanoflagellates in the Antarctic sea ice (Moorthi *et al.* 2009). In this study in the Ross Sea, mixotrophic nanoflagellates contributed 5–10% of phototrophic nanoflagellates and 3–15% of bacterivorous nanoflagellates in ice cores. Cryptophytes and *Pyramimonas gelidicola* ingest both bacteria and DOC in saline Antarctic lakes (Laybourn-Parry *et al.* 2005) and may do so in the sea ice.

Chlorophyll *a* and rates of primary production are shown in Table 4.1. Assimilation numbers or chlorophyll-specific rates of photosynthesis are widely available for sea ice algae in contrast to the paucity of data for surface ice environments (see Section 3.2.4). A range of values are shown in Table 4.3. Assimilation numbers will of course vary in relation to season and levels of PAR, and with depth in the ice, which may show an inverse trend (Lizotte and Sullivan 1992). A review of assimila-

tion numbers of marine phytoplankton provides a mean of 1.38 ± 0.89 g C (g Chl a)$^{-1}$ h^{-1} (Markager *et al.* 1999). The efficiencies exhibited by sea ice photosynthetic communities appear comparable. As one might expect there are clear diurnal variations in the rates of primary production that correlate closely with daily changes in the levels of under-ice irradiance (Figure 4.7). Highest productivity occurs around midday and is lowest at midnight. Interannual variations are also clearly apparent. For example, between 2002 and 2004 investigations in October at different latitudes provided maximum rates of primary production of between 2.6 mg C m^{-2} h^{-1} (in 2002) and 5.9 mg C m^{-2} h^{-1} (in 2004) (McMinn *et al.* 2007).

Another very clear feature of sea ice communities is the spatial heterogeneity of the communities, both autotrophic and heterotrophic. This is a feature that has been widely noted in many investigations. As we have already seen in this chapter and in Section 1.4.1 the sea ice is not a homogeneous medium.

Table 4.3 Assimilation numbers or chlorophyll *a*-specific rates of photosynthesis for Arctic and Antarctic sea ice algal communities.

Location	Assimilation number (µg C (µg Chl a)$^{-1}$ h^{-1})	Source
Off Labrador pack ice winter	0.08–0.85	Irwin 1990
Hudson Bay fast ice in spring	April 1.8 ± 0.9	Gosselin *et al.* 1986
	May 5.2 ± 2.1	
Weddell Sea and western Antarctic Peninsula in autumn and winter	0.04–8.6	Lizotte & Sullivan 1992
McMurdo fast ice in summer	0.2–1.5	Stoecker *et al.* 2000
McMurdo Sound congelation ice in spring	0.09–1.6	Palmisano *et al.* 1987

Table 4.4 Percentage of total primary production in the Antarctic seasonal ice zone contributed by sea ice algae.

				Month			
	Oct	Nov	Dec	Jan	Feb	Mar	Apr
Mean %	10.3	8.8	3.4	1.6	1.0	1.3	1.5

Mean values derived from data for the Weddell Sea, the Indian Ocean, the Pacific Ocean, the Ross Sea, and the Bellinghausen-Amundsen Sea.
From Lizotte (2001).

The fact that the period of primary production in the sea ice occurs at a phase in the annual cycle when productivity in the water column is extremely low or lacking is crucially important to the marine ecosystem. This pulse of productivity provides food for metazoans like krill and copepods. Estimates of the contribution of the sea ice autotrophic community to overall production in the Southern Ocean indicate that while the contribution to total production is usually low, it is the timing of that production that is of fundamental importance (Table 4.4).

4.1.3.2 Bacterial production

Sea ice typically has high concentrations of dissolved organic carbon relative to the underlying water. Usually, concentrations of organic matter in water are 10–100-fold lower than in the ice (Thomas and Papadimitriou 2003). For example in pack ice in the Amundsen Sea (Antarctic) DOC ranged from 50 to 100 μ M C (Thomas *et al.* 1998), and up to 398 μ M C in the Weddell Sea (Dumont *et al.* 2009), and in the Fram Strait (Arctic) in multi-year pack ice from less than 100 to 700 μ M C (Thomas *et al.* 1995). There are high levels of spatial heterogeneity in its concentration. In the western Weddell Sea DOC concentrations ranged from 45 to 669 μ M C kg^{-1} (Underwood *et al.* 2010). In many cases a good correlation exists between DOC and chlorophyll *a* concentration (Riedel *et al.* 2007; Underwood *et al.* 2010), though the relationship is not always apparent. Dumont *et al.* (2009) found high DOC associated with high chlorophyll *a* levels in the lower ice layers, but in other layers high DOC corresponded to low concentrations of chlorophyll *a*. Similarly in the Fram Strait there was no correlation between DOC and chlorophyll *a* (Thomas *et al.* 1995). The bulk of sea ice DOC derives from sea ice diatoms that exude considerable quantities of extracellular polymeric substances (EPS) (see Section 1.4.2). Bacteria also exude DOC but their contribution to the pool is small.

The pool of DOC provides an energy source for heterotrophic bacteria (see Section 1.4.2). While a portion of this material is used as a substrate for bacterial digestion, part of the EPS form into exopolymer particles (EPs) by a number of processes. Biological activity forms capsules and sheets but most EPs are created abiotically by coagulation of DOM and colloidal precursors. These particles are larger than 1 μm and are retained by 1μm pore filters. Concentrations of EPs in sea ice are high, for example ranging from 10.2 to 260 $\times$ 10^6 L^{-1} with an area of 3.4–92.1 cm^2 L^{-1} (Meiners *et al.* 2004). EPs are well colonized by bacteria both on their surfaces and within the particle. Attachment to EPs undoubtedly enhances the capacity for production. In open water in both the sea and in lakes bacteria attached to particles achieve higher rates of growth. EPs in the sea ice contribute to particle fluxes in the open water (called 'marine snow') when the ice melts.

The bacteria of some sea ice communities appear much more morphologically diverse than those seen in seawater. On the Mackenzie ice shelf and the Resolute Passage (Arctic) and Prydz Bay (Antarctic) fast ice, the bacterial community contained significant numbers of large filamentous bacteria (Laurion *et al.* 1995; Riedel *et al.* 2007; Paterson and Laybourn-Parry 2011). However, in other locations there is no apparent variation in morphology or cell volume (Smith and Clement 1990).

Based on cultured isolates subjected to 16S rRNA sequence analysis the pack ice bacterial community of the Chukchi Sea possessed a close relationship with marine psychrophilic bacteria from the genera *Alteromonas*, *Colwellia*, *Glaciecola*, *Octadecabacter*, *Psuedoaltermonas*, and *Shewanella* within the Proteobacteria, and *Cytophaga*, *Flavobacterium*, *Gelidibacter*, and *Polarbacter* in the Cytophaga-Flexib acter-Bacteroides (Junge *et al.* 2002). The data suggest low genetic diversity compared to the water column, which is a characteristic of extreme environments. Among the genera identified, the Flavobacteriaceae of the Cytophaga-Flexibacter-Bacteriodes group are known to show a preference for attachment to surfaces. The sea ice offers an environment with numerous attachment sites, for example exopolymer particles. The majority of bacteria are unculturable, so that the application of culture-independent molecular techniques, such as fluorescence *in situ* hybridization (FISH), provides a more comprehensive picture of diversity. A study that compared Antarctic (Weddell Sea) with Arctic (Fram Strait) pack ice bacteria diversity used both cultures and culture-independent techniques (FISH) (Brinkmeyer *et al.* 2003). Among the Arctic samples 50% were Gammaproteobacteria, and in the Antarctic samples 36%. The Gammaproteobacteria included the genera *Shewanella*, *Psychromonas*, *Colwellia*, *Glaciecola*, *Pseudoalteromonas*, *Methylophaga*, *Psychrobacter*, *Marinobacter*, *Oceanospirillum*, and *Halomonas*. Twenty-five per cent of rest of the community were Alphaproteobacteria (*Roseobacter*, *Sphingomonas*) and 25% were identified as belonging to the Cytophaga-Flavobacterium group including *Flavobacterium*, *Cytophaga*, *Actinobacteria*, and *Polaribacter*. A higher diversity of phylotypes in the Arctic samples, which included limnetic phylotypes, suggested a terrestrial influence on the sea ice bacterial communities of the Fram Strait.

However, a strong similarity between the phylotypes seen in samples from both polar locations suggests similar selective mechanisms, despite differences in their quantitative contribution to the communities as revealed by FISH.

Abundances of bacteria in sea ice are shown in Table 4.5. While in many cases data are reported as numbers per litre of melted ice, allowing useful comparisons between sites and seasons, there is much less consistency in relation to the reporting of production data (Table 4.5). One of the inevitable problems with a three-dimensional solid medium such as ice and soil is that data can be expressed per litre (or gram if soil) or on a volumetric (m^{-3}) or areal basis (m^{-2}). In aquatic environments data are conventionally expressed per unit volume, which facilitates comparisons.

There is considerable variation in the reported abundances of bacteria found in sea ice. In part this undoubtedly reflects the heterogeneity of the environment, but will also be related to the time of year the study was conducted and the nature of the ice (fast ice, pack ice, multi-year ice). The crucial point, however, is that within extreme environments, providing there is liquid water, even as a film, bacterial production or growth occurs throughout the year, including winter. This is also apparent from Section 3.2.4.

As indicated above, there is often a correlation between the concentration of DOC and chlorophyll *a* and primary production. It follows that there are

Table 4.5 Bacterial abundances and production in Arctic and Antarctic Sea ice.

Site	Bacterial abundance $\times 10^9$ L^{-1}	Bacterial production (μg C L^{-1} h^{-1})	Source
Arctic			
Mackenzie Ice Shelf	0.2–2.7	–	Riedel *et al.* (2007)
Laptev Sea	1.2–4.8	–	Krembs and Engel (2001)
Barrow Sea	20	–	Smith *et al.* (1989)
Northern Baltic Sea	0.07–0.27	0.02–0.12[2]	Haecky and Andersson (1999)
Antarctic			
Bellinghausen Sea	0.5–4.8	–	Meiners *et al.* (2004)
Weddell Sea	2.8 max	0.01–0.4	Grossmann and Dieckmann (1994)
Weddell Sea	10	0.008–0.11[3]	Helmke and Weyland (1995)
Western Antarctic Peninsula	–	0.1–0.79[4]	Kottmeier and Sullivan (1987)
Prydz Bay	0.03–0.21	0–1.4[5]	Paterson and Laybourn-Parry (2011)

1, pmol leucine L^{-1} h^{-1}; 2, μmol C m^{-2} h^{-1}; 3, mg C m^{-3} h^{-1}; 4, mg C m^{-2} h^{-1}; 5, study covered ice formation to ice breakup.

often clear couplings between primary production and bacterial biomass and production (e.g. Kottmeier and Sullivan 1987; Haecky and Andersson 1999). In the Baltic Sea bacterial production was equivalent to 0.7–11.9% of primary production between January and April, with the highest percentage occurring in January (Haecky and Andersson 1999), while in the Resolute Passage bacterial production represented 3% of net primary production during April to June (Smith and Clement 1990). Within the sea ice of the McMurdo Sound bacterial production was 9% of primary production (Kottmeier *et al.* 1987). These are relatively low proportions when compared with values reported from the open waters of the Southern Ocean, which can range between 33% and 75% (Ducklow 1983). However, this feature of bacterial production, representing only a small proportion of primary production, appears typical of extreme low temperature habitats. For example in cryoconite holes on glaciers bacterial production is considerably less than 1% of primary production (see Section 3.2.4). Undoubtedly temperature is an important limiting factor, as the sea ice usually offers a rich supply of substrates in terms of DOC, and nutrients essential for growth (nitrogen and phosphorus) are not generally limiting.

4.1.3.3 *Heterotrophic protozoa*

Wherever there are bacteria, their major predators, the heterotrophic nanoflagellates (HNAN), are also found. In aquatic environments the HNAN are the major predators of heterotrophic bacteria. There is little information on the species diversity of the HNAN in the sea ice. Choanoflagellates common in

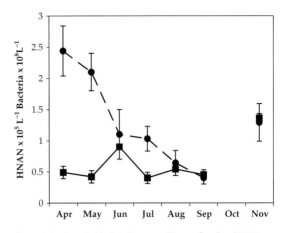

Figure 4.8 Bacteria (dots) and heterotrophic nanoflagellate (HNAN, squares) abundances over an annual cycle in the sea ice of Prydz Bay. Data from Paterson and Laybourn-Parry (submitted).

the water column are also seen within the sea ice, for example *Acanthocorbis* and *Diaphanoeca* (Ikävalko and Gradinger 1997; Paterson and Laybourn-Parry 2011). Other reported HNAN species include *Cryothecomonas* spp., *Leucocryptos*, *Paraphysomonas*, *Rhynchomonas*, and *Telonema* (Garrison and Buck 1991; Archer *et al.* 1996; Ikävalko and Gradinger 1997; Meiners *et al.* 2002; Paterson and Laybourn-Parry 2011). There is much more information on the abundance of HNAN (Table 4.6). There are few studies that cover the winter, but the limited data available show that bacteria and their predators are present and functional during winter darkness (Kaartokallio 2004; Paterson and Laybourn-Parry 2011) (Figure 4.8). While there was no correlation between HNAN numbers and their prey, nonetheless during winter

Table 4.6 Heterotrophic nanoflagellate (HNAN) and ciliate abundances in sea ice.

Site	HNAN × 10^4 L^{-1}	Ciliates (L^{-1})	Source
Baltic Sea fast ice (W)	2.2	5000*	Kaartokallio (2004)
McMurdo fast ice (Su)	1.4–5.7	1000–5000	Stoecker *et al.* (1993)
Gulf of Finland pack ice (S)	1.0–55.0	–	Meiners *et al.* (2002)
Weddell Sea pack ice (S)	18.1–22.4	545	Garrison and Buck (1991)
Prydz Bay fast ice (W/S/Su/A)	2.1–29.8	0–3100	Paterson and Laybourn-Parry (2011)

S, spring; Su, summer; A, autumn; W, winter. * *Mesodinium rubrum.*

darkness when photosynthesis is curtailed through lack of PAR, heterotrophic carbon cycling continues.

The grazing impact of HNAN on the bacterial community has been assessed in the Arctic (Resolute Passage) in land-fast ice (Laurion *et al.* 1995). Grazing rates decreased between April and May, with ingestion rates ranging between <3 and 64 bacteria HNAN^{-1}h^{-1}. These ingestion rates were insufficient to meet the observed growth of the HNAN community. However, some HNAN species, particularly choanoflagellates, are capable of exploiting DOC of a range of molecular weights (Marchant and Scott 1993; Laybourn-Parry *et al.* 2005). The missing carbon to support growth may well have been derived from the ingestion of DOC. These data suggest that patterns of HNAN grazing changed during the ice algae bloom. Bacterivory was high during the early stages of the bloom, but decreased as the bloom reached it peak, at which point the HNAN may have been exploiting the DOC pool to a greater degree.

We still have much to learn about the grazing activities of heterotrophic protozoa in sea ice. Research on the microbial communities of extreme Antarctic lakes has demonstrated that nutritional versatility is fundamental to survival (Laybourn-Parry *et al.* 2005). It is very likely that the heterotrophic protozoan communities of sea ice possess this characteristic. Another complicating factor when assessing grazing is that not all the bacterial community is available as potential food. Where there are large filamentous bacteria within the community, these will be too large for most HNAN to ingest. Similarly, attached bacteria, which may represent a significant portion of the community in sea ice, are not accessible to all HNAN. Some species of HNAN, however, are adapted to feeding effectively on attached bacteria, for example *Bodo* (Caron 1987). Protozoa practice selectivity in feeding. We have previously assumed that simple single-celled organisms are incapable of selective feeding, but this is not the case. Some bacteria are more nutritionally attractive than others, and even when these bacteria are offered as one-fifth of the total bacterial mélange the protozoans will continue to select these in preference to other species (Thurman *et al.* 2010). Grazing rates of heterotrophic and mixotrophic protozoans in a wide range of aquatic habitats have

been measured by following the ingestion of heat-killed bacteria that have been stained with a fluorescent dye, known as fluorescently labelled bacteria (FLBs). The bacteria consumed during incubations can be seen inside the protozoan cell under fluorescent microscopy. This technique assumes no selectivity and that the protozoans do not discriminate against dead cells that have a chemically changed cell surface covered by a dye. It is now possible to insert different coloured fluorescent proteins into live bacterial cells that are clearly visible inside a protozoan when they are ingested. The bacterial cell surfaces are not changed in any way. The advent of this technique allows mixtures of different coloured bacteria to be offered in feeding experiments and the possibility of selective feeding to be elucidated. Moreover, grazing rates on these live bacteria are usually higher than on FLBs, so we may have been underestimating grazing rates.

Heterotrophic dinoflagellates are a conspicuous component of the protozoan community of sea ice (Ikävalko and Gradinger 1997). The majority are larger than nanoflagellates, some being in excess of 20 µm in length (see Table 1.1 and Figure 1.4). Sea ice dinoflagellates include members of the genera *Gymnodinium*, *Polykrikos*, *Gyrodinium*, and *Protoperidinium* and are reported in concentrations of up to 2.2×10^4L^{-1} in McMurdo Sound fast ice (Stoecker *et al.* 1993) and 0.4×10^3L^1 in Weddell Sea pack ice (Garrison and Buck 1991). They contributed up to 15% of the heterotrophic protozoan biomass in McMurdo Sound, 25% in Prydz Bay, and 11.5% in the Weddell Sea (Garrison and Buck 1991; Stoecker *et al.* 1993; Archer *et al.* 1996). Little is known about the physiology of sea ice heterotrophic dinoflagellates. They are likely to feed on bacteria, nanoflagellates, and small diatoms and possibly DOC and may be significant in carbon cycling.

Dinoflagellates appear to outnumber ciliates in the land-fast ice (Stoecker *et al.* 1993; Paterson and Laybourn-Parry 2011), whereas in pack ice they were outnumbered by ciliates (Garrison and Buck 1991). Typical concentrations of ciliates are shown in Table 4.6 and the species listed in Table 4.7. The mixotrophic species *Mesodinium rubrum* (see Figure 1.2A) is a conspicuous element and can reach high concentrations (see Table 4.6). This is an interesting

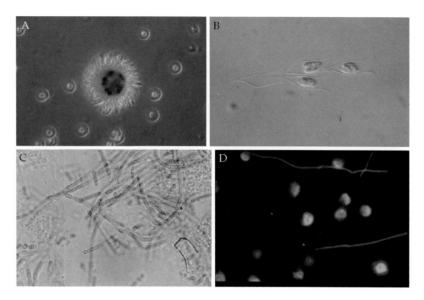

Plate 1 (A) *Mesodinium rubrum*: a ciliate fixed in Lugol's iodine, where the dark staining areas are the endosymbiotic cryptophyte; (B) *Dinobryon*: a colonial phytoflagellate (photo courtesy of W. Vincent); (C) Cyanobacteria *Leptolyngbya fragalis* (photo courtesy of Jeff Johansen and Mark Schneegurt (www-cyanosite.bio.purdue.edu)); (D) cryptophytes under epifluorescence microscopy. *M. rubrum*, *Dinobryon*, and cryptophytes are all mixotrophic species. See also Figure 1.2, page 5.

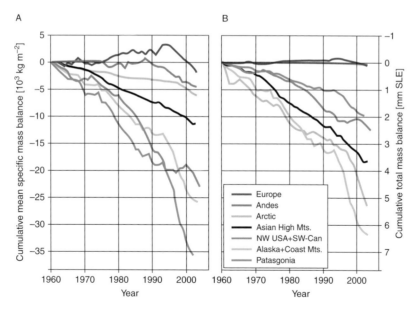

Plate 2 The decline in terrestrial ice mass in large regions (Dyurgerov and Meier 2005) as shown by (A) the cumulative mass balance (in terms of tonnes of ice lost per square metre) and (B) the cumulative effect on sea level (in terms of millimetres). The relative strength of the response of climate change by different regions is clearly shown in (A), with Patagonia loosing most ice in terms of mass per unit area. By contrast, the glaciers of Alaska and the surrounding mountains have the largest effect on sea level rise (B). SLE, sea level equivalent. From *Climate Change 2007: the Physical Science Basis. Working Group I Contribution to the Fourth Assessment Report of the Intergovernmental Panel on Climate Change*, Figure 4.15, Cambridge University Press. See also Figure 1.6, page 11.

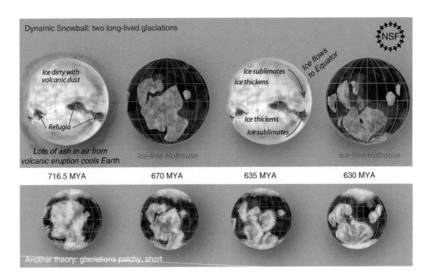

Plate 3 A schematic representation of the growth of ice on Snowball Earth, and the subsequent melting of the snowball as greenhouse gases (principally CO_2) build up in the atmosphere. Reproduced with the permission of Pearson Education. See also Figure 1.7, page 11.

Plate 4 Pack ice in the Southern Ocean. See also Figure 1.9, page 13.

Plate 5 Aerial view of Taylor Valley, one of the McMurdo Dry Valleys in Antarctica. These cold polar glaciers have characteristic cliffs around their margins, which vary from a few to 20 m or more. See also Figure 1.16, page 19.

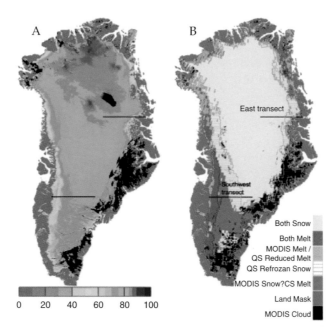

Plate 6 Extent of melting of surface snow and ice on the Greenland Ice Sheet during July 2007, as determined by satellite remote sensing. (A) The albedo of the surface is near 100% wherever there is cold snow, meaning that all incoming solar radiation is reflected back into the atmosphere. As the snow starts to melt, the albedo drops, and when all the snow has melted to reveal the underlying grey ice, the albedo drops even further, down to as little as 30%. MODIS refers to data collected by Moderate-resolution Imaging Spectrodiometry, and QS refers to melt calculated from the Daily QuikSCAT Melt Special Product algorithm. More detail can be found in Hall et al (2009). From Hall *et al.* (2009). See also Figure 1.19, page 22.

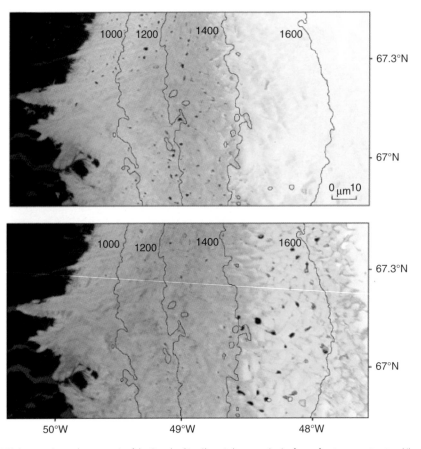

Plate 7 Supraglacial lakes near the southwest margin of the Greenland Ice Sheet. Lakes range in size from a few tens or metres to a kilometre or so across. The upper satellite image was taken on 11 June, whereas the lower satellite image was taken on 16 July. Note how, in the latter, there are more supraglacial lakes within the region from which the overlying snow pack has melted. The black lines denote the altitude of the ice surface. From Sundall *et al.* (2009). See also Figure 1.22, page 24.

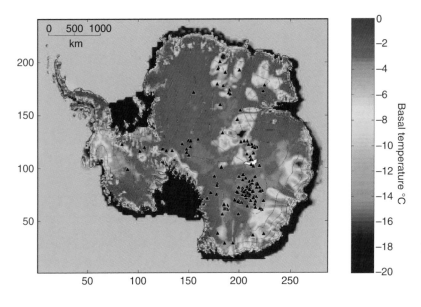

Plate 8 Basal temperature of the Antarctic Ice Sheet derived from the thermo-mechanical ice sheet model. Redrawn from Siegert *et al.* (2006). See also Figure 1.25, page 27.

Plate 9 Red snow created by snow algae on the Signy Island ice cap, martime Antarctic. See also Figure 1.28, page 35.

Plate 10 Ward Hunt Ice Shelf (Canadian Arctic) showing parallel, elongated surface lakes. Photo courtesy of W. F. Vincent. See also Figure 3.1, page 48.

Plate 11 A cyanobacterial mat on the bottom of the littoral region of an Arctic lake. Photo courtesy of W. F. Vincent. See also Figure 3.2, page 49.

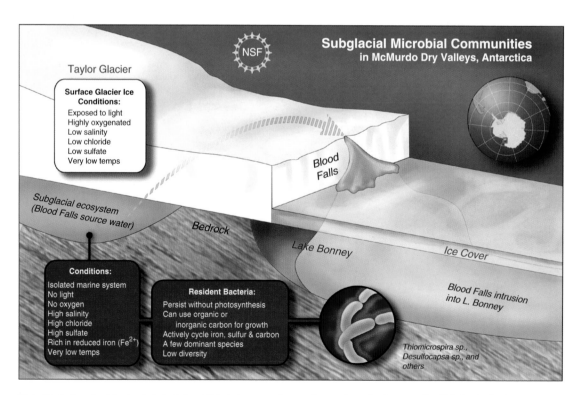

Plate 12 A schematic of the chemical, physical, and biological processes that contribute to the outflow known as Blood Falls from Taylor Glacier, Antarctica. Illustration by Zina Deretsky, NSF. See also Figure 5.1, page 108.

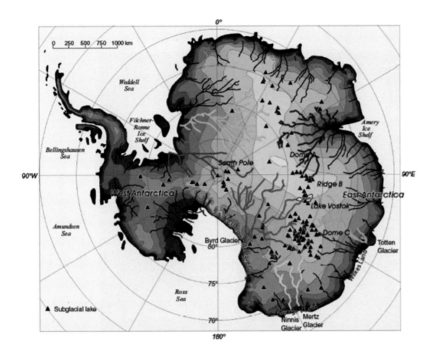

Plate 13 The major arterial tunnels that are thought to carry large volumes of water from the base of the Antarctic Ice Sheet towards the margins. The position of some of the many subglacial lakes found beneath Antarctica are also shown. From Siegert *et al.* (2009), and with the permission of John Wiley and Sons. See also Figure 5.5, page 112.

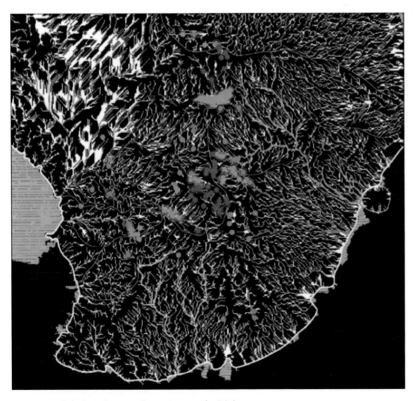

● Lakes that are downstream of <2 lakes
◉ Lakes that are downstream of 2-5 lakes
◉ Lakes that are downstream of >5 lakes

▬ Lakes that are upstream of>5 known lakes
▬ Lakes that are upstream of 2-5 lakes
▬ Lakes that are upstream of <2 lakes

Plate 14 Fine detail of the subglacial drainage system that could exist beneath the southeast sector of the Antarctic Ice Sheet. Note that the putative drainage system connects many subglacial lakes. From Siegert *et al.* (2009), and with the permission of John Wiley and Sons. See also Figure 5.6, page 113.

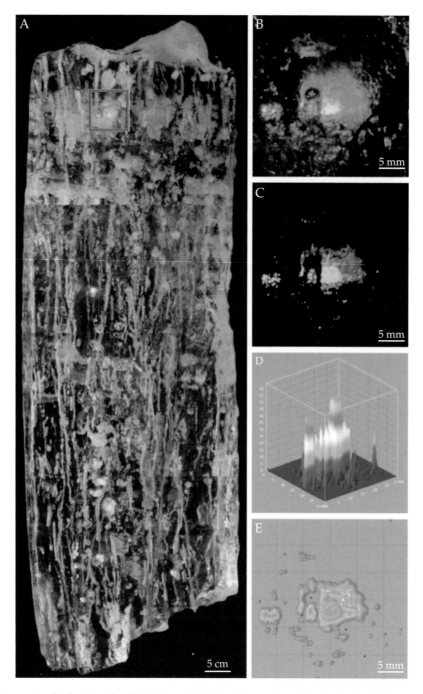

Plate 15 (A) A 65 cm section of ice from the surface of Lake Untersee, Antarctica, with cryoconite particles at various depths. Fluorescence was stimulated using a 532 nm laser. (B) The reflected laser light from the target at the top of the core in (A) (see box near top of image), and its fluorescence (C). (D, E) The processed images in 'light cone' and isobar formats, respectively. From Storrie-Lombardi and Sattler (2009), with permission of Mary Ann Liebert Inc. See also Figure 7.5, page 136.

species, which is widespread in the sea and estuaries and can form red tides when it blooms. It is a species complex and was thought to contain an endosymbiotic cryptophycean relying entirely on its endosymbiont for its nutrition (Lindholm 1985). However, it is now known that at least some of the members of this complex feed on cryptophytes and sequester their plastids, which they then use for photosynthesis (Gustafson *et al.* 2000). *M. rubrum* has a wide salinity tolerance. This is evident from the fact that it has successfully colonized the marine-derived saline lakes of the Vestfold Hills in Antarctica, where it achieves concentrations of $100\,000\,L^{-1}$, and is found in lakes with salinities as low as 4‰ and as high as 62‰ (27% higher than seawater) (Perriss *et al.* 1995; Laybourn-Parry *et al.* 2002). Thus it is particularly well adapted to living in the sea ice where salinities can vary considerably. Ciliates exploit a wide range of food sources. Smaller ciliates, e.g. scuticociliates, feed voraciously on bacteria; larger species feed on both autotrophic and heterotrophic nanoflagellates, small dinoflagellates, and small diatoms. At low temperatures their growth rates are likely to be low with long doubling times.

Another protozoa group reported from the sea ice is that of the heliozoans (Ikävalka and Gradinger 1997). These are members of the Sub-Phylum Sarcodina, which also includes amoebae and foraminiferans. As their name implies they resemble stylized suns. They feed on nanoflagellates, small ciliates, and metazoans. When a large prey item is captured by an individual, others heliozoans may fuse with the feeding individual, separating again at the end of the digestion process (Patterson and Hausmann 1981). Naked amoebae are also likely to be common in the sea ice as they are usually associated with surfaces because they need to be attached to feed; surfaces of various types are common in the sea ice environment. They are virtually impossible to see in fixed samples and can only really be assessed from cultured material. To date they are a group that has been overlooked.

Foraminifera, which like the naked amoebae and Heliozoa are members of the protozoan Sub-Phylum Sarcodina, are also reported for sea ice studies. They are conspicuous because they are large-shelled protists possessing compartmented shells. They have been reported as reaching concentrations of 47.8×10^3 individuals m^{-2} during October to

Table 4.7 Ciliate species and groups seen in pack ice and land-fast ice.

Genera	Weddell Sea pack ice[1]	McMurdo Sound fast ice[2]	Prydz Bay fast ice[3]
Oligotrich spp.	*		
Strombidium spp.		*	*
Rimostrombidium			*
Pelagiostrombidium			*
Strobilidium		*	
Tontonia			*
Laboea		*	
Tintinnid spp.		*	
Tintinnopsis			*
Cymatcylis			*
Didinium		*	
Monodinium			*
Mesodinium rubrum	*	*	*
Lacrymaria		*	
Spriroprorodon		*	
Scuticociliates		*	*
Hypotrich spp.	*		
Aspidisca		*	
Euplotes	*	*	*

1, Garrison and Buck (1991); 2, Stoecker *et al.* (1993); 3, Paterson and Laybourn-Parry (2011).

December in the Weddell Sea (Schnack-Schiel *et al.* 2000). Foraminifera have not been found in Arctic sea ice. It is believed that they are incorporated into developing ice in the Antarctic, and then grow inside the brine channel network. Their absence from the Arctic sea ice may have something to do with the way ice forms in the Arctic, making it difficult for this protist to be entrained into developing ice (Gradinger 1999).

4.1.3.4 Viruses

As indicated in Section 1.4.2, there is very little information on viruses in the sea ice. This is somewhat surprising given that there has been significant research conducted on them and their role in both seawater and freshwater ecosystems. Certainly viruses are abundant in polar seas (Marchant *et al.* 2000), where they can play a very important role in carbon cycling (see Section 1.2.1). Phage–host systems have been isolated from sea ice samples taken from Arctic sea ice (Borriss *et al.* 2003). The bacterial hosts showed greatest similarity (based on 16S rDNA analysis) to *Flavobacterium hibernum*, *Shewanella frigidimarina*, and *Colwellia psychreryth-raea*, all of which are psychrophilic with extremely good growth at low temperatures. They also showed a wide salinity tolerance, growing in media that were twice the salinity of seawater and in media that were 25% seawater. Their phages were tailed and had double-stranded DNA, making them members of the phage families *Siphoviridae* and *Myoviridae*. The phages were host specific, that is they were not able to infect a wide range of hosts, but were restricted to one species.

Concentrations of virus-like particles in Antarctic (Prydz Bay) fast ice ranged from $2.9 \times 10^7 \, L^{-1}$ in winter (July) to $1.9 \times 10^9 \, L^{-1}$ in summer (late November) and in the brine from $4.6 \times 10^7 \, L^{-1}$ in winter to $3.1 \times 10^9 \, L^{-1}$ in summer (Paterson and Laybourn-Parry 2011), while in pack ice in the Ross Sea abundances reached higher levels, ranging from 6.3×10^6 to 1.19×10^8 viruses mL^{-1} in summer and 10^6 to 10^8 viruses mL^{-1} in autumn (Gowing *et al.* 2002, 2004). During a spring ice algal bloom in the fast ice near Resolute in the Canadian Arctic viruses reached between $9.0 \times 10^9 \, L^{-1}$ and $1.3 \times 10^{11} \, L^{-1}$ (Maranger *et al.* 1994). Here the sea ice viral concentrations far exceeded those in the underlying water ($1.1 \times 10^9 \, L^{-1}$).

The underlying water in Prydz Bay, however, had viral concentrations not dissimilar to those in the ice, where maximum values varied between $3.5 \times 10^9 \, L^{-1}$ and $2.1 \times 10^{10} \, L^{-1}$ (Pearce *et al.* 2007; Paterson and Laybourn-Parry 2011). In the Ross Sea up to 18% of the viruses were large with a capsid diameter exceeding 110 nm. These were likely to infect eukaryotes like nanoflagellates (Gowing 2003). The concentrations of viruses in any given habitat depends on the abundance of hosts and the capability of those hosts to produce more viruses, which in turn depends on the availability of suitable carbon substrate and nutrients (nitrogen and phosphorus) to support bacterial production. The number of viral particles in the environment, which is what the counts given above show, is also a function of the degree of lysis versus lysogeny that prevails. A further factor is that viruses in the free environment decay. Light has an important impact on decay rates, even when ultraviolet (UV) wavelengths are excluded (Wommack *et al.* 1996; Wilhelm *et al.* 2003). In winter when there is no light, decay is likely to be very slow, and in summer the high attenuation of light in ice, particularly snow-covered ice, will also mediate against rapid decay.

Virus to bacterial ratios (VBRs) are widely quoted to provide an indication of the number of host cells available for potential viral infection. A review of this parameter in a range of freshwater and marine ecosystems indicates that in freshwater the VBR ranges between 4.9 and 77.5 (mean 20–25) and in marine waters ranges between 0.38 and 53.8 (mean 1–5) (Maranger and Bird 1995). In the Arctic sea ice VBRs ranged from around 10 up to 72 in spring (Maranger *et al.* 1994), and a similar high VBR was noted in the Ross Sea (range 0.7–119) (Gowing *et al.* 2004), while in Prydz Bay the VBR ranged between 0.4 and 26.1 over a year with no obvious seasonal pattern (Paterson and Laybourn-Parry 2011). VBRs vary considerably across ecosystems, for example in saline marine-derived Antarctic lakes VBRs were high, ranging from 30.4 to 96.7 (Madan *et al.* 2005). Studies in ice environments like Arctic glacier cryoconites showed a VBR range between 7.4 and 26.6, while in extremely unproductive Antarctic freshwater lakes the VBR is typically below 5 (Säwström *et al.* 2008). Generally VBRs in extreme environments appears to be low, related to low levels of bacterial

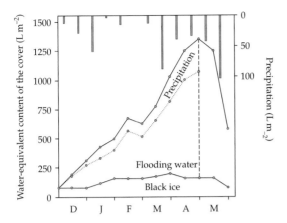

Figure 4.12 Time-related changes in the water equivalent content of the ice cover of Lake Redó. The origin of the water is indicted for the growth period. The grey vertical lines indicate snow deposition, the top solid line indicates precipitation, and the dotted line below it indicates flooding water over black ice (bottom curve). Note that in May some rainfall occurred. Redrawn from Felip *et al.* (1999).

water equivalent volume of the ice cover was accounted for by flooding (up to 78%) (Felip *et al.* 1999).

In Lake Redó, chlorophyll *a* in the slush layer deceased from 5 to 0.3 µgL⁻¹ between January and the end of April. The initial high values coincided with similar levels in the water column. During the phase of ice cover there was a clear species succession. Algae including desmids and chlorococcales found at the beginning quickly disappeared. Volvocales, chrysophytes, cryptophytes, and heterotrophic nanoflagellates occurred in higher densities in the slush than in the underlying water column. Cryptophytes, chrysophytes, and hetrotrophic flagellates dominated the growth phase of the ice cover with volvocales, dinoflagellates, and ciliates being more abundant during the cover formation. This is attributed to their abundance in the lake water. Some of the rotifers from the water column colonized the ice sporadically during its growth phase. These included *Kelliocottia*, *Polyarthra*, and *Asplanchna* spp. (Felip *et al.* 1999). There was a high diversity of species from a range of taxonomic groups within the ice, as shown in Table 4.10. Cyanobacteria also occurred (Table 4.10), but unlike the Dry Valleys lakes they did not dominate the biomass. The majority of the species found in the ice will have colonized from the flooding lake waters.

However, a few of the ciliates such as *Urostyla*, *Lacrymaria*, and *Dilpetus* noted in the ice of Lake Redó and lakes in the Tyrolian Alps (Felip *et al.* 1995) are not planktonic species but are characteristic of the benthos or bottom of the lake. They may have become entrained in the flooding water and colonized the ice slush layer. The much higher biological diversity found in the ice covers of these lakes reflects less harsh conditions in comparison with the Antarctic Dry Valleys.

In common with sea ice, the bacterial communities of Lake Redó (Spanish Pyrenees), Gossenköllesee, and Schwarzee ob Sölden (Tyrolian Alps) exhibited considerable morphological diversity, with cocci, filaments, and short rods. Their abundances and rates of productivity are shown in Table 4.11. Bacterial concentrations are overall lower in lake ice than in sea ice (see Table 4.5). In lake ice values up to 16×10^8 L⁻¹ were recorded, while in sea ice the range is from 0.007 to 48×10^8 L⁻¹. This is related to the availability of carbon substrate and nutrients required to support bacterial growth. Both sea ice and lake ice have good supplies of nitrogen and phosphorus from underlying and flooding waters, though concentrations are higher in the sea ice. The sea ice also has higher levels of DOC, much of it derived from the dense diatom communities.

There are few data on primary production in lake ice, the only information relates to Lake Redó and Lake Bonney (Dry Valleys). In the former, rates of photosynthesis in May ranged from 0.277 to 0.405 µg C L⁻¹ h⁻¹ (Felip *et al.* 1995). The assimilation numbers or chlorophyll *a* specific rates of photosynthesis ranged from 0.98 to 1.4 µg C (µg Chl *a*)⁻¹ h⁻¹. Rates of primary production in the ice exceeded those in the water (range 0.002–0.115 µg C L⁻¹ h⁻¹). This is what one would anticipate, given that penetration of PAR to the water column was attenuated by the ice cover. Primary production in Lake Bonney ranged up to 0.325 µg C L⁻¹ h⁻¹, which is comparable to the rates reported from Lake Redó. However, in Lake Bonney this was mainly achieved by Cyanobacteria, whereas in Lake Redó the autotrophic community was diverse.

As melt occurs, pools form on the surface of the slush layer; in some ways these water bodies are analogous to the pools encountered on the surface of ice shelves (see Section 3.1). However, they differ

Table 4.10 Taxomomic diversity in Lake Redó ice cover during its annual cycle.

Taxonomic group	Species or genera
Cyanobacteria	*Synechocystis* sp.,*Chroococcus* spp.
Chlorophytes:	*Chloromonas* spp., *Chloromonas infirma*, *Chloromonas modesta*, *Chlamydomonas* spp., *Chlamydomonas nivalis*,
Volvocales	*Pteromonas* sp., *Provasoliella* sp., *Dysmorphococcus variables*,*Dictyospaerium* sp., *Oocystis parva*, *Spaerocystis*
Chlorococcales	*schroeteri*,*Cosmarium* sp., *Botryococcus braunii*, *Ankistrodesmus fusiformis*
Chrysophyceae:	*Chromulina* spp., *Chryoscoccus* sp., *Chrysolykos skujae*, *Kephyrion planctonicum*, *Ochromonas* spp., *Psuedokephyrion* sp.,
Autotrophic forms	*Dinobryon cylindricum*,*Spumella* spp., *Oikomonas termo*
Heterotrophic forms	
Chryptophytes	*Chroomonas acuta*, *Cryptomonas marsonii*, *Cryptomonas ovata*, *Rhodomonas minuta*
Dinoflagellates	*Gymnodinium* spp., *Gymnodinium cnecoides*, *Gymnodinium* cf. *lantzschii*, *Gymnodinium austriacum*, *Peridinium incospicuum*
Choanoflagellates	*Desmarella* sp.
Other heterotrophic flagellates	*Bodonid*-like species, amoeboid flagellate species
Ciliates	*Askenasia* spp., *Holophrya* spp., *Mesodinium pulex*, *Urotricha furcata*, *Urotricha pelagica*, *Rimostrombidium* cf. *humile*, *Strombidium* sp., *Pelagiostrombidium fallax*, *Uronema* sp., *Sphaerophyra* sp., *Dileptus* sp., *Lacrymaria* sp., *Rhopalophyra* sp.

Data from Felip *et al.* (1999).

Table 4.11 Bacterial abundance and rates of production based on leucine incorporation into protein for Lake Redó (2240 m altitude), Gossenköllesee (2415 m altitude), and Schwarzee ob Sölden (2799 m altitude).

Lake	Bacterial abundance × 10⁸ L⁻¹	Bacterial production (pmol leucine L⁻¹ h⁻¹)
Redó	0.73–6.40	1.3–6.0
Gossenköllesee	2.10–6.30	3.6–54.3
Schwarzee ob Sölden	2.6–16.00	1.3–211.5

Data from Felip *et al.* (1995).

in their biota. Ice shelf lakes are dominated by Cyanobacteria, whereas the pools on alpine lakes contain much the same assemblage as the slush although they also harbour a few species not derived from the underlying lake waters, notably *Chlamydomonas nivalis*, other *Chlamydomonas* sp., and a species of *Chromulina*. *C. nivalis* is one of the so-called snow algae and is common on the surface of snow (see Section 2.2.1). At the end of the winter the slush layers and the pools are enriched by aerial deposition of pollen, leaves, and other organic debris. Their bacterial concentrations are comparable to those in the slush layer and range from 2.10 to 4.80×10^8 L⁻¹ and their productivity between 16.5 and 28.2 pmol leucine L⁻¹ h⁻¹. There is only one datum for primary production in surface pools for Lake Redó in May when the photosynthetic rate was 0.06 μg C L⁻¹ h⁻¹ (Felip *et al.* 1995).

Molecular analysis of the bacterial communities of the ice profile and lake waters in winter revealed variation between snow, slush, and lake water. Probes specific for the α-, β-, and Gammaproteobacteria and Cytophaga-Flavobacterium showed differences between the various habitats, although in most cases communities were dominated by members of the Betaproteobacteria (Alfreider *et al.* 1996).

Limited data suggest that there may be differences between ice structure on lakes at altitude and lakes close to sea level. This may be related to depth and size of the lake and the amount of snowfall. For example, on shallow, fluvial Lac Saint-Pierre on the St Lawrence River in Canada ice thickness was between 62 and 97cm overlying water up to 1.26 m in depth. Here the lower ice was dense, clear, and transparent, while the upper layer was composed of opaque white ice. Snow cover was limited because of prevailing wind patterns. The ice contained vertical channels about 1–5 cm wide in its lower 30 cm during March. Algae. when they were present. were most abundant at the ice–water interface and in channels, and included up to 55 taxa of Cyanobacteria, diatoms, ciliates, and rotifers. While some of the taxa were similar to those recorded in Lake Redó (Table 4.10), there was a dominance of diatoms like *Diatoma* spp., *Fragilaria* spp., and *Melosira*. Chlorophyll *a* concentrations were high, reaching 169 µgL^{-1}. When melt occurs, these populations are likely to seed the underlying phytoplankton, as has been described for sea ice diatom communities (Frenette *et al.* 2008).

Lake Baikal is the world's large largest lake by volume and its deepest (average depth 730m). It is an ancient lake formed in a rift valley and today has an area of 31 500 km². It contains many endemic species. The lake possesses ice up to 110 cm thick that is transparent, but during winter it develops an intricate network of cracks, some of which form into crevasses. In April the ice loses its transparency and melts. It is described as falling apart like needles (http://www.irkutsk.org/baikal/). This is similar to the annual 2 m thick ice seen on coastal Antarctic lakes, like those of the Vestfold Hills. Lake Baikal has been a site of limnological research for many years, but sadly many of the data are published in Russian journals and are not easily accessible to the rest of the scientific world.

Diatoms have been described from the ice–water interface and in the interstitial water-filled cracks within the ice of Lake Baikal. They have a complex life cycle of moving between ice and benthos. *Aulacoseira baicalensis* was studied using underwater video recording. It grows within the ice body, forming aggregations that hang down into the water. The algae secrete mucus that creates these aggregations. They can detach from the ice and float freely in the water, eventually sinking to the lake bottom. Another closely related diatom species *Aulacoseira skvortzowii* forms similar finer aggregates (Bondarenko *et al.* 2006).

Clearly lake ice represents another largely seasonal habitat in the cryosphere. To date it has attracted relatively little attention, however studies on lake water columns in winter are not that common either. Ice has a major impact on the water column beneath it. It removes wind-driven turbulence and can have a profound effect on the light climate of the underlying water, particularly in spring. The reduction in PAR transmission through ice is further enhanced when the ice carries a snow cover. As indicated above, ice can impose an important effect on the lake water temperature. It has been argued that there is a pressing need to study lakes and their ice covers in winter. Ice is an integral part of the functioning and dynamics of lakes and what happens in winter affects what follows in spring and summer (Salonen *et al.* 2009).

CHAPTER 5

Subglacial environments

5.1 Introduction

A key requirement for microbial life in subglacial environments is the presence of water, so it is important for biologists to understand the types of drainage systems that occur under glaciers. This is because these drainage systems control the likely Eh and pH, the supply of freshly ground minerals for energy and nutrient sources, and the locus of any labile organic carbon (Tranter *et al.* 2005). It surprises many that water is often present at the beds of many glaciers and ice caps (Hooke 1989), and over large areas (~50%) of the beds of the Greenland and Antarctic Ice Sheets (Bell 2008). This is because the small amount of geothermal heat rising through the bedrock is often just enough to melt and maintain a thin layer of water at the interface between the bedrock and the ice (Paterson 1994). Those areas of the bed where water is present are termed warm-based areas, which contrast with cold-based areas of the bed where the glacier is frozen to the bedrock. Usually, small, thin glaciers at high latitudes are cold based. By contrast, smaller glaciers at lower latitudes and larger glaciers, ice caps, and ice sheets often contain appreciable areas of warm beds. Most larger glaciers at higher latitudes have cold-based margins and some warm-based ice in their interior. Theses glaciers are often referred to as polythermal based. It is currently believed that warm beds host a spectrum of microbial life (Tranter *et al.* 2005), although direct evidence is sparse.

The beds of smaller, warm-based glaciers are fed by surface meltwater, which usually flows in large supraglacial channels until these channels intercept crevasses or moulins (Benn and Evans 2010) (see Figure 1.23). The latter are large tubes that extend from the surface to the bed. Moulins are almost always associated with their own large supraglacial channel. They were originally crevasses that were prevented from closing by the water flow from the supraglacial channel, which melts the moulin walls by frictional heating and the warming associated with water falling from high to low elevation (since potential energy must be converted to kinetic energy). This melting counteracts the tendency for ice to anneal any holes and fractures by deformation during ice flow. Frictional heating also melts tunnels or channels at the glacier bed (Kamb 1987). These channels, with dimensions of width and height of the order of metres, transport most of the water from the subglacial drainage system to the terminus, and form an arborescent network called the channelized drainage system. The large diurnal pulses of ice melt from the surface maintain the channelized drainage system, which would otherwise revert to a so-called distributed drainage system by deformation of the overlying ice, which tends to collapse the tunnels and channels (Kamb *et al.* 1985). The distributed drainage system underlies the snow-covered areas of the glacier, where there is a smaller and more constant seepage of melt from the snow-covered surface, arising from the higher albedo of ice and the porous nature of a wet snow pack (Richards *et al.* 1996). Snow melt either drains to the bed via crevasses and moulins beneath the snow cover, or drains onto the snow-free glacier and contributes to supraglacial flow. The distributed drainage system flanks the channelized drainage system beneath snow-free areas of the glacier surface, and drains into the channelized drainage system via a channel marginal zone, rather similar to the hyporeic zone of surface rivers and streams (Hubbard *et al.* 1995). Both the channelized drainage system and the channel marginal zone flood during high discharge, and drain during falling discharge. Hence, there is the potential for some ingress of atmospheric gases into the channelized drainage system during low discharge, particularly near the glacier terminus. By contrast, the distributed drainage system is almost always full of water and under high pressure because of the overburden pressure, which relentlessly tries to squeeze the water pockets shut. The principal subglacial route by which water drains from valley glaciers becomes increasingly the channelized drainage system as the melt season progresses and the snow line moves up the glacier (Richards *et al.* 1996).

The type of subglacial drainage system model detailed above was developed for glaciers with hard beds, meaning those with no basal till. Many glaciers flow over basal till, and are called soft bedded (Benn and Evans 2010). Water can flow through melted basal till, particularly under the ice streams (Tulaczyk *et al.* 2000). Flow and water pressures within the till are likely to be low because of the relatively high clay and silt content that is typical of till. It is probable that many glaciers have beds with areas that are hard, soft, and intermediate, or patchy. However, it is believed that all bed types have rapid water drainage through a channelized drainage system that is either melted into the overlying ice (called R- or Rothlisberger channels) or melted into the underlying frozen till or that have eroded into the underlying bedrock, particularly in limestone areas—the latter being known as N- or Nye channels (Paterson 1994).

The subglacial drainage system structure of larger polythermal glaciers, ice caps, and ice sheets is currently an area of active research and debate (Bell 2008; Bartholomew *et al.* 2010). Essentially, these types of ice masses are all believed to have a distributed drainage system or to permit the flow of water through the till and underlying bedrock. They are likely to have a channelized drainage system near the glacier terminus if surface meltwater can penetrate to the bed. The cold-based margins that these larger systems at higher latitudes often have may temporarily dam the first meltwaters reaching the bed in the vicinity of the terminus. There are often outburst floods when meltwater first hydrofractures the cold ice (Skidmore and Sharp 1999). Thereafter, flow at the bed quickly melts large channels, which promote the development, if only temporarily, of a channelized drainage system beneath the margin. However, the rest of the warm bed is occupied by a distributed drainage system and/or water-laden till. The source of meltwater beneath the ice sheets is due to geothermal heating and internal deformation of the basal ice. The waters have no contact with the atmosphere, and so only carry the oxygen content of the parent ice. Hence, given appropriate compounds to oxidize, such as Fe(II) and Mn(II) minerals, sulphides, and organic matter, there is great potential for sub-ice sheet environments to become suboxic or anoxic (Wadham *et al.* 2008, 2010).

The margins of the Greenland Ice Sheet are characterized by the presence of numerous and large supraglacial lakes, particularly in the south (Box and Ski 2007) (see Figure 1.22). The lakes have dimensions from 100m to several kilometres, and they may be many tens of metres deep. Hence, they carry considerable volumes of water. Any fracturing of ice in the lake floor due to glacier motion results in water penetrating the crack. The water may refreeze and warm the surrounding ice, but subsequent glacier motion serves to increase the probability that water-filled cracks will eventually penetrate down to the bed, and hence promote lake drainage. Flowing water melts ice, so the flow becomes greater and floods the bed, causing temporary speed up of the ice margin in the vicinity of the lake (Das *et al.* 2008). However, the rapid flow of water to the bed results in any high pressure distributed drainage system being melted out into a low pressure channelized drainage system, so acting against more rapid motion (Sundal *et al.* 2011).

The Antarctic Ice Sheet is underlain by hundreds of subglacial lakes, bodies of ponded water, often contained within topographic low spots within the bed (Siegert *et al.* 2005). They have dimensions of hundreds of meters to tens of kilometers and depths of tens to hundreds of meters. It seems likely that many of the lakes are at least periodically connected and, if this is so, there will be a spectrum of hard and soft bed hydrological environments connecting them (Tulaczyk and Hossainzadeh 2011). Some of these flow paths will be hydrologically connected at all times, while some will at least partially freeze between connection events (Wingham *et al.* 2006). This means that there will a wide spectrum of water residence times—from days through potentially thousands of years—and a wide spread of water-laden environments—from continuously open, water-filled channels through to water-filled, fine basal debris through which water flow is very slow. These environments potentially offer niches for a broad spectrum of microbes, and hence contribute to the biodiversity of microbial life under ice sheets (Tulaczyk and Hossainzadeh 2011).

Other major contributors to microbial diversity under ice are volcanoes and geothermal hotspots. Volcanoes exist beneath the Vatnajökull Ice Cap, Iceland (Gaidos *et al.* 2008), and are thought to occur beneath the Greenland Ice Sheet (Fahnestock *et al.* 2001). Hydrothermal sources are believed to occur within Lake Vostok, beneath the Antarctic Ice Sheet (Souchez *et al.* 2004). The hot waters emanating from these systems are likely to contain specialist microbes (Bulat *et al.* 2004; Gaidos *et al.* 2008).

The types of biogeochemical reactions that microbes undertake in subglacial environments depend on the source of the water and the flow path the water follows (Tranter 2004; Tranter *et al.* 2005). Waters flowing from the surface to the bed clearly carry relatively large quantities of oxygen to the bed, and oxidizing conditions prevail along the predominantly channelized flow paths. Waters formed by regelation, geothermal heating, and internal deformation of ice beneath the ice sheets inject relatively small quantities of oxygen to the glacier bed, and hence waters flowing in distributed drainage systems, particularly beneath larger ice masses, become increasingly anoxic over time (Tranter *et al.* 2002a; Wadham *et al.* 2008). Microbially mediated reactions involve oxidation and reduction, so-called REDOX reactions, and in many ways, the REDOX reactions in glacial sediments are similar to those that occur in freshwater lake and river sediments. The most likely reactions to occur are those that liberate most energy. So, where present, organic matter is oxidized first by oxygen (Equation 5.1), and the CO_2 released is used to chemically dissolve minerals in the bedrock, a process known as carbonation. There is a limited body of evidence which suggests that microbial oxidation of bedrock organic matter occurs (Wadham

et al. 2008), and, if this is the case, carbonation as a consequence of microbial respiration may occur in debris-rich environments, such as the distributed drainage system and the channel marginal zone (Tranter 2004).

$$\underset{\text{organic matter}}{C_{org}(s)} + O_2(aq) + H_2O(l) \qquad (5.1)$$

$$\leftrightarrow H^+(aq) + HCO_{3^-}(aq)$$

Sulphide minerals are a ubiquitous component of most types of rock at the Earth's surface. Glacial comminution is very efficient at liberating sulphide minerals from within bedrock, since the surface area to mass ratio of glacial debris is so high relative to the bedrock. Hence, one of the most dominant REDOX reactions in subglacial environments is sulphide oxidation (Equation 5.2). Sulphide oxidation occurs predominantly in debris-rich environments, where comminuted bedrock is first in contact with water (Raiswell *et al.* 2009). It is microbially mediated, occurring several orders of magnitude faster than in sterile systems (Sharp *et al.* 1999). It consumes oxygen, driving down the pO_2 of the water. The oxidation of sulphides preferentially dissolves carbonates, rather than silicates, because the rate of carbonate dissolution is orders of magnitude faster (Tranter *et al.* 2002b). However, carbonates may become exhausted over time or the water may become saturated with respect to carbonate in some sub-ice stream environments. Thus silicate weathering predominates in sub-ice sheet environments (Skidmore *et al.* 2010; Wadham *et al.* 2010).

$$\underset{\text{pyrite}}{4FeS_2(s)} + \underset{\text{bedrock carbonate}}{16Ca_{1-x}(Mg_x)CO_3(s)} \qquad (5.2)$$

$$+15O_2(aq) + 14H_2O(l)$$

$$\leftrightarrow 16(1-x)Ca^{2+}(aq) + 16xMg^{2+}(aq)$$

$$+16HCO_{3^-}(aq) + 8SO_4^{2-}(aq) + \underset{\text{ferric oxyhydroxide}}{4Fe(OH)_3(s)}$$

Sulphide oxidation rapidly removes oxygen from subglacial meltwaters and produces anoxia (Bottrell and Tranter 2002), and in these anoxic environments, oxidizing agents such as Fe(III) are used for sulphide oxidation (Equation 5.3). Sources of Fe(III) include the products of the oxidation of pyrite, magnetite, and haematite.

$$FeS_2(s) + 14Fe^{3+}(aq) + 8H_2O(l) \leftrightarrow$$
$$15Fe^{2+}(aq) + 2SO_4^{2-}(aq) + 16H^+(aq) \qquad (5.3)$$

The resupply of glacial systems with new or recent organic matter is limited to that in-washed from the glacier surface, such as algae, insects, and animal faeces, or from overridden soils during glacier advance. By contrast, the supply of old organic matter from comminuted rocks is plentiful. Given the thermodynamic instability of organic matter in the presence of O_2 or SO_4^{2-}, it seems likely that microbes will have evolved to colonize subglacial environments and utilize bedrock organic matter as an energy source. The first support for this assertion was from stable isotope analysis at Finsterwalderbreen, a small polythermal-based glacier on Svalbard, which has shale as a significant component of its bedrock (Wadham *et al.* 2004). The $\delta^{18}O$-SO_4 of water upwelling from subglacial sediments is very enriched in ^{34}S, which suggests that cyclical sulphate reduction and oxidation has been occurring. The $\delta^{13}C$ of dissolved inorganic carbon (DIC) is negative, consistent with the assertion that organic matter has been oxidized. Mass balance calculations suggest that the most probable source of organic matter is bedrock organic matter. Hence, sectors of the bed at Finsterwalderbreen appear to be anoxic, where a significant geochemical weathering reaction is sulphate reduction linked to oxidation of bedrock organic matter (Equation 5.4).

$$2CH_2O9(S) + SO_4^{2-}(aq)$$
bedrock organic matter
$$\leftrightarrow 2HCO_3^-(aq) + H_2S(aq) \qquad (5.4)$$

It is possible that methanogenesis (Equation 5.5) occurs under certain ice masses, since methanogens have been isolated from subglacial debris (Skidmore *et al.* 2000; Wadham *et al.* 2008).

$$2CH_2O(s) \leftrightarrow CO_2(aq) + CH_4(aq) \qquad (5.5)$$

Glacial runoff is a supplier of potentially bioavailable nitrogen and phosphorus to the oceans (Hodson *et al.* 2004; Hood and Scott 2008), and a major source of a very different type of dissolved organic carbon (DOC) to the refractive forms found in riverine runoff from terrestrial sources. Glacial DOC is labile and is readily available for heterotrophic bacterial production in high latitude coastal waters (Anesio *et al.* 2009; Hood *et al.* 2009; Fellman *et al.* 2010). DOC concentrations in glacial runoff are typically 0.2–0.5mgCL^{-1}, lower than that in riverine runoff by a factor of two to four (Hood *et al.* 2009). The protein- and acetate-rich DOC that glacial microbes produce (Hood *et al.* 2009) is more labile than the relatively refractive, aromatic- and lignin-rich DOC that results from the degradation of terrestrial organic matter (Holmes 2008), and is readily utilizable by heterotrophic bacteria in coastal waters (Berggren *et al.* 2009; Fellman *et al.* 2010). The DOC originates both from the surface and subglacial environments (Hood and Berner 2009; Bhatia *et al.* 2010). The former results from both autochthonous production (Anesio *et al.* 2009) and from allochthonous sources (Stibal *et al.* 2008), while the latter results from the oxidation of ancient overridden organic matter (Willerslev *et al.* 2007; Bhatia *et al.* 2010).

5.2 Biology of subglacial environments

Unlike snow, supraglacial environments, and sea and lake ice, subglacial environments lack light. In the dark photosynthesis is not possible, so heterotrophic, chemoheterotrophic, and chemoautotrophic organisms are found in these extreme environments. Chemoautotrophic bacteria use reduced inorganic substrates for the reductive assimilation of carbon dioxide and as a source of energy. These bacteria can exploit the reduction of H_2, NH_3, NO_2^-, H_2S, or Fe_2^+ as a source of energy. Aerobic forms use oxygen as a terminal electron acceptor, while anaerobic *Archaea* species can use inorganic sulphur as a terminal electron acceptor. Chemoheterotrophs exploit organic substrates as a source of carbon and energy. Among this group, sulphate reducers use

H_2 for energy but require an organic substrate to achieve growth.

Conditions beneath glaciers vary considerably depending on the type of glacier. The substrate over which the glacier rides is important in determining the carbon supply. For example, kerogens are the organic chemical compounds that constitute the organic matter in sedimentary rocks, so glaciers overriding these rocks will have a ready supply of organic carbon to support bacterial growth. In contrast, glaciers overriding hard volcanic rocks will lack an endogenous ready supply of carbon and will probably support much more limited communities. As outlined above, warm-based glaciers will have liquid water, essential to support life, while in cold-based glaciers the basal sediment may be frozen for most of the time, precluding active metabolism. The availability of oxygen can also vary, with hypoxia and anoxia occurring under some glaciers. Thus, this is the domain of aerobic as well as facultative and obligate anaerobes. It has to be said that at present our knowledge of life under glaciers is limited and as far as functional dynamics are concerned, we are still to a large extent in the realm of speculation.

5.2.1 Wet-based glaciers

While information is limited, glaciers from as far apart as New Zealand, Alaska, and Ellesmere Island in Canada have been investigated from a biological perspective. Polythermal glaciers possess warm ice (at 0°C) in their interior, where the ice is thickest, and cold ice at their margins, where the ice is thinner (see Section 5.1). There is accumulating evidence that biological activity plays an important role in the dissolution and oxidation of minerals at the base of glaciers when liquid water is present. Molecular biology plays a crucial role in attempts to unravel the composition of the bacterial communities and the role they may play in biogeochemical processes. This approach, in conjunction with geochemical and stable isotope investigations, provides a multidisciplinary understanding of one of the most extreme icy environments. Unlike supraglacial environments (see Chapter 3), the base of a glacier is dark, precluding photosynthesis, and cold and entirely the domain of prokaryotes. The eukaryotic

Protozoa and metazoans found in surface glacial environments cannot survive in habitats beneath glaciers.

It is an accepted fact that the majority of bacteria (~99%) are not culturable. In other words, when one cultures and isolates bacteria from the natural environment what you get is not really representative of the community. However, we now have the molecular tools to extract and amplify 16S rRNA genes from environmental samples. A study on the Franz Joseph and Fox Glaciers in the South Island of New Zealand adopted a culturing approach, but in this case they were able to culture between 3% and 82% of the bacteria sampled (Foght *et al.* 2004). This undoubtedly reflects the simplicity of the community. Bacterial cell concentrations ranged between 2 and 7×10^6 g^{-1} dry weight of sediment. This value is low when compared to bacterial abundances in cryoconite holes on glacier surfaces, where concentrations ranged between 3.0 and 9.9×10^8 g^{-1} (see Section 3.2.3), and lake ice, where values ranged from 0.73 to 16×10^8 L^{-1} (see Section 4.2). Nevertheless, it shows that a functional microbial community of bacteria may be present below glaciers.

The two New Zealand glacier isolates showed a high proportion of Betaproteobacteria (11 of 23 isolates). Among these were bacteria that clustered with *Polaromonas vacuolata* and *Rhodoferax antarcticus*, or with clones obtained from permanently cold environments. Bacteria were cultured on media specific for particular physiological groups. In the Fox Glacier as well as aerobic and microaerophilic heterotrophs, nitrate-reducing bacteria, nitrogen-fixing bacteria, and ferric iron-reducing bacteria were present. The same groups occurred in the Franz Joseph Glacier, with the exception of nitrogen-fixing bacteria, which were undetectable. Interestingly, the majority of the isolates proved to be psychrotolerant rather than psychrophiles and all were cultured under aerobic conditions (Foght *et al.* 2004). In contrast, most of the isolates from an Arctic glacier proved to be psychrophiles (Skidmore *et al.* 2000), which throws up an interesting biogeographical question. The two New Zealand glaciers override glacial till and debris containing temperate vegetation at the margins. The total concentrations of carbon and nitrogen were lower than has been reported in glaciers elsewhere in Switzerland and the Arctic, implying greater microbial oxidation of DOC and reduction of nitrate.

Limited evidence to date suggests that the Betaproteobacteria are a major component of glacier microbial assemblages. A dominance of Betaproteobacteria was evident from clone libraries created from subglacial samples derived from the Bench Glacier in Alaska and the John Evans Glacier in the Canadian High Arctic (Skidmore *et al.* 2005) (Table 5.1) and they also dominated the clone library derived from the Fox and Franz Joseph Glaciers in New Zealand (Foght *et al.* 2004). Subglacial sediments recovered from beneath the Kamb Ice Stream under the West Antarctic Ice Sheet had abundant Betaproteobacteria among cultured isolates (Lanoil *et al.* 2009). Betaproteobacteria appear common on glacier surfaces too. Cryoconite holes on glacier surfaces in the McMurdo Dry Valleys have a dominance of Betaproteobacteria in their ice covers, though the cryoconite sediment had a high percentage of Cytophaga-Flavobacterium-Bacteriodes (Foreman *et al.* 2007). However, a comparative study of bacterial communities in basal ice and in supraglacial, subglacial, and proglacial meltwaters from the John Evans Glacier showed that the 16S rRNA genes amplified from these samples were different. Subglacial water, basal ice, and sediment communities were distinct from those found in supraglacial meltwater and proglacial sediments. These differences were attributed to variations in the physico/chemical environment (Bhatia *et al.* 2006).

In total 127 clones were obtained from the Bench and John Evans Glaciers. While the Betaproteobacteria dominated the clone libraries from each glacier, there were differences between the two locations (Table 5.1). The John Evans Glacier had no Deltaproteobacteria or Epsilonproteobacteria, whereas these groups constituted 3.2% and 10.6% of the clones obtained from the Bench Glacier. However, the John Evans Glacier had 33 clones belonging to the Cytophaga-Flavobacterium-Bacteriodes group, while the Bench Glacier had few (Table 5.1). Many of the sequences associated with the Betaproteobacteria from the Bench Glacier included *Polaromonas vacuolata*, *Rhodoferax* spp., and *Comamonas* spp. Relatives of *Thiobacillus* (Betaproteobacteria) that are involved in acidic iron oxidation and/or neutrophilic sulphur oxidation were present among the clones in significant levels. In the John Evans Glacier clone library *Comamonas* spp. were very well represented (44% of the Betaproteobacteria clones), however *Thiobacillus* were less well represented. Overall, the John Evans Glacier possessed a greater diversity than that derived from the Bench Glacier (Skidmore *et al.* 2005). *Comamonas* and its relatives were predomi-

Table 5.1 Distribution of clones in the clone libraries for the John Evans Glacier (133 clones) and the Bench Glacier (94 clones) for major taxonomic groups.

Phylogenetic group	John Evans Glacier		Bench Glacier	
	No. of clones	% of clones	No. of clones	% of clones
Alphaprotoebacteria	10	7.5	4	4.3
Betaproteobacteria	74	55.6	64	68.1
Gammaproteobacteria	2	1.5	10	10.6
Deltaproteobacteria	0	0	3	3.2
Epsilonproteobacteria	0	0	10	10.6
Cytophaga-Flavobacterium-Bacteriodes	33	24.8	1	1.1
Holophaga-Acidobacteria	2	1.5	1	1.1
Spirocheata	0	0	1	1.1
Planctomycetales	2	1.5	0	0
Actinobacteria	7	5.3	0	0
Verrucomicrobium	3	2.3	0	0

Data from Skidmore *et al.* (2005).

Wynn-Williams 1996), and lake bed sediments may hold records of the past climate change that resulted in the glaciation of Antarctica (Barrett 1999). Some 350 subglacial lakes are known to exist beneath Antarctica (Wright and Siegert 2011).

Some 50% of the bed of the Antarctic Ice Sheet is believed to be warm based (Bell 2008), and there is great potential for an extensive subglacial drainage system to exist. Figure 5.5 shows where the major hydrological flow paths might exist, and Figure 5.6 shows the finer resolution of the potential drainage system in the vicinity of Lake Vostok (Siegert *et al.* 2007). The potential for many of the lakes to be interconnected is very apparent. Current research

aims to establish the nature of the connectivity of the lakes (Tulaczyk and Hossainzadeh 2011).

The chemistry of the lake waters has yet to be determined from direct measurements, but estimates have been made from the chemical composition of the accretion ice (Siegert *et al.* 2003), refrozen lake water, at the base of the Vostok Ice Core (Figure 5.7). The reconstruction of the water composition is controversial, and there may be several different waters types within the lake (De Angelis *et al.* 2004). The waters may be derived entirely from meteoric ice melt and so be very dilute, or they may be quite concentrated and derived from hydrothermal-type fluids (Table 5.3). Waters sampled directly from the

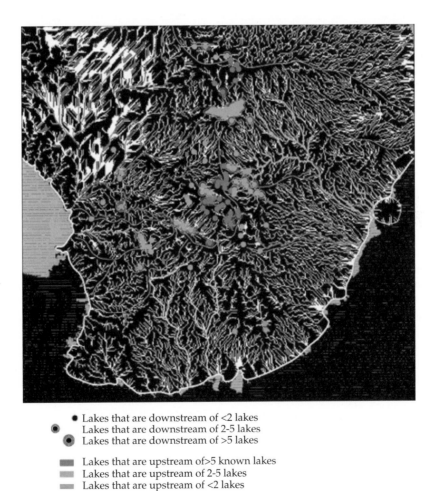

- Lakes that are downstream of <2 lakes
- Lakes that are downstream of 2-5 lakes
- Lakes that are downstream of >5 lakes

- Lakes that are upstream of >5 known lakes
- Lakes that are upstream of 2-5 lakes
- Lakes that are upstream of <2 lakes

Figure 5.6 Fine detail of the subglacial drainage system that could exist beneath the southeast sector of the Antarctic Ice Sheet. Note that the putative drainage system connects many subglacial lakes. From Siegert *et al.* (2007), with the permission of John Wiley and Sons. See also Plate 14.

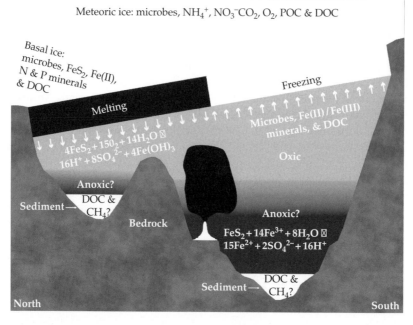

Figure 5.7 Chemical compounds that are likely to be delivered to subglacial Lake Vostok, which lies some 4 km beneath the centre of the Antarctic Ice Sheet. The chemical compounds shown may be used by microbes to sustain life in the subglacial lake. DOC, dissolved organic carbon; POC, particulate organic carbon. From Christner *et al.* (2008), and with the permission of Springer Science+Business Media B.V.

bed of the ice sheet from beneath the Kamb and Bindschadler Ice Streams are very concentrated, testimony to the higher rock:water ratio and the long residence time of water in the sediments (Table 5.4). It is likely that there is extensive microbial activity in these water-saturated sediments (Skidmore *et al.* 2010). However, relatively little is known about the nature of the microbes at present. What is clear is that there are many different REDOX species that can be used by any microbes present as a source of metabolic energy, as shown in Figure 5.7. These species can be derived from the melting of ice, from sediments, and from any hydrothermal activity that might exist in the lake (Christner *et al.* 2006).

Table 5.3 Minimum and maximum estimates of the chemical composition of Lake Vostok water.

Estimate	Na^+	Ca^{2+}	Mg^{2+}	Cl^-	SO_4^{2-}	HCO_3^-
Minimum	0.005	0.001	0.002	0.002	0.002	0.001
Maximum	11	3.3	2.3	8.3	4.6	3.4

Units are mmol L^{-1}.
After Siegert *et al.* (2003).

Much of the speculation about what the Lake Vostok ecosystem might be like comes from studies on the accretion ice at the base of the core. As yet the lake waters have not been penetrated. The upper 3310m of the core is glacial ice that provides an

Table 5.4 The concentration of major ions in waters sampled from beneath the Kamb and Bindschadler Ice Streams, KIS and BIS respectively.

Location	pH	Na^+	K^+	Ca^{2+}	Mg^{2+}	Cl^-	SO_4^{2-}	HCO_3^-
KIS	6.5	20.0	0.6	6.4	3.1	1.1	17.0	4.1
BIS	6.5	35.0	0.7	9.0	8.6	2.0	31.0	7.5

Units are mmol L^{-1} (except for pH).
After Skidmore *et al.* (2010).

environmental record for four complete ice age cycles. The section of the core between 3310 and 3539m is transitional between glacial and accretion ice, while ice below 3539 m it is accretion ice that is refrozen lake water accreted to the bottom of glacial ice. Accretion ice is likely to contain samples of the lake water biota and provide information on important aspects of lake biogeochemistry, such as concentrations of DOC and major ions (Priscu *et al.* 1999). One of the major challenges in identifying organisms in accretion ice is ensuring that what one is looking at is from the lake and not a contaminant from the drilling process or subsequent treatment of the ice core. Clearly stringent precautions are necessary and there has been debate as to whether the bacteria identified are indeed derived from the lake (Christner et al. 2001; Bulat *et al.* 2004). Considerable effort is being invested in developing technologies that will eliminate any chance of contamination when drilling (see Section 7.4). The subglacial lakes of Antarctica represent a unique environment that has been isolated from the atmosphere for millions of years, and are consequently of considerable biological interest.

The lake waters originate from the overlying ice sheet that melts near the shoreline of the lake and at the ice–water interface in the northern portion of the lake. The dissolved oxygen concentration is thought to be around 50 times higher than air-equilibrated water. The waters should be supersaturated with dissolved gases in equilibrium with clathrate in the water column. This is an air hydrate composed of crystallized water (ice) molecules that form a rigid lattice of cages with most of the cages containing a molecule of air (McKay *et al.* 2003). Major ions and organic carbon all decrease with depth in the accretion ice. Depth in the ice represents a proxy for increasing distance from the shoreline, suggesting that there may be greater potential for biological activity in the shallow shore water of the lake. Organic carbon concentration in ice that accreted from lake water in a shallow embayment in the southwestern portion of the lake (type I accretion ice) was 65 µmol L^{-1}, and in ice formed over the deep water (type II accretion ice) it was 35 µmol L^{-1}. This organic carbon originates from the lake sediments or the overlying ice sheet (Christner *et al.* 2006). Incubations of melted accretion ice with

^{14}C-labelled acetate and glucose resulted in the release of ^{14}C-CO_2, indicating metabolically active cells, albeit at extremely low levels (Karl *et al.* 1999). These data, together with molecular analysis of the organisms entrapped in accretion ice, provide compelling evidence that life exists in subglacial lakes.

Bacterial cell density is two to seven-fold higher in the accretion ice than in the overlying glacial ice. Highest cell numbers in the accretion ice reached 380± 53 cells mL^{-1} of melted ice (Christner *et al.* 2006). The bacteria showed a spectrum of cell sizes with small coccoid cells (0.1–0.4 µm) constituting about half of the community, while the other half was made up of rods and vibrios (Karl *et al.* 1999). Based on isolated cultures and the amplification of small-subunit rDNA molecules, molecular analysis revealed phylotypes that belong to the Alphaproteobacteria, Betaproteobacteria, and Gammaproteobacteria, the Firmicutes, and the Actinobacteria and Bacteriodes lineages (Priscu *et al.* 1999; Christner *et al.* 2001; Bulat *et al.* 2004; Christner *et al.* 2006) (Figure 5.8). The closest relatives of some of the phylotypes provide clues to the functional dynamics of the lake water community. For example, among the Betaproteobacteria clone, sequences were closely related (96–97%) to aerobic methylotrophic species in the genera *Methylobacillus* and *Methylophilus*. These bacteria use C-1 compounds as a substrate (e.g. methanol, formate, carbon monoxide) and carbon assimilation is via the ribose monophosphate pathway. Thus there may be niches for methylotrophy in Lake Vostok (Christner *et al.* 2006). The thermophilic facultative chemolithoautotrophic bacterium *Hydrogenophilus thermoluteolus* (Betaproteobacteria) has been identified from accretion ice. It has only previously been found in hot springs and may use hydrogen for energy and carbon dioxide as a carbon source (Bulat *et al.* 2004). Based on the evidence to date, Priscu *et al.* (2008) have produced a diagram showing potential biogeochemical pathways for chemoautotrophic bacteria that may operate in the surface waters of Lake Vostok (Figure 5.9).

Inevitably there has been much speculation that Lake Vostok and other Antarctic subglacial lakes may possess a unique microbiota, given their period of isolation from the rest of the biological world. Priscu *et al.* (2008) provide a good review of the issue. They argue that if the accretion

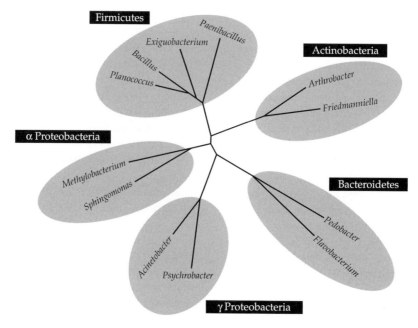

Figure 5.8 Genera most frequently recovered from glacial ice and subglacial environments based on the phylogeny of the 16S rRNA molecule. The source environments are very diverse but have in common that they are permanently frozen. From Priscu *et al.* (2008) with permission of Oxford University Press.

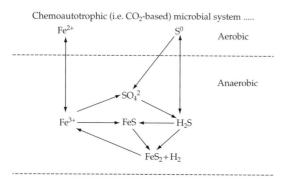

....supplying organic C for use by heterotrophs:
$(CH_2O)n + X_{oxidised} \rightarrow CO_2 + X_{reduced}$
Where $X_{oxidised}$ may be O_2, NO_3^-, SO_4^{2-}, S^0, $Fe3^+$

Figure 5.9 Potential biogeochemical pathway that may be important in the surface waters of Lake Vostok based on 16S rDNA sequence data. The pathway involves chemoautotrophic fixation of CO_2 using iron and sulphur as both electron donors and electron acceptors. The pathway also shows the potential transformations that may occur across an aerobic–anaerobic boundary. Organic carbon produced by chemoautotrophic metabolism could then fuel heterotrophic metabolism in Lake Vostok. From Priscu *et al.* (2008) with permission of Oxford University Press.

ice bacteria are representative of the lake water community, they do not represent an evolutionary distinct biota. Lake Vostok has been isolated for >15 million years, which is really a very small period of time in the context of the Earth's history and the species divergence of the prokaryotes. The possibility exists that glacial meltwater that enters the lake may form a lens overlying the main water body. If this is the case the microorganisms found in the accretion ice may not have spent much time in the lake water column before being incorporated into the accretion ice. In other words, what is seen in the accretion ice may have originated from the glacial ice. Alternatively, the biota in the main body of the water column may have originated from the basal sediments. If that is the case then there may have been time for some degree of evolutionary divergence that will not be revealed until we have samples of the water column.

Two ecological scenarios have been proposed for Lake Vostok (Christner *et al.* 2006). Firstly, organic carbon derived from the sediments and/or the overlying ice sheet supports an exclusively heterotrophic community. Measurable aerobic respiration is apparent in some of the accretion ice samples. The cell concentration in the near-surface water column is estimated to lie between 140 and 770 cells

mL^{-1}. From the calculated rate of organic carbon input, the predicted cell numbers, and a cell carbon content of 10 fg cell^{-1}, and assuming no abiotic sinks for organic carbon within the lake, a positive organic carbon flux would provide a heterotrophic bacterial community with 0.49–3.8× 10^{-4} g organic C (g cell C)$^{-1}$ h^{-1}. Even if all the organic carbon proves a suitable substrate, this would not meet the theoretical carbon demand of the community. It would be adequate to support maintenance but not net growth. Christner *et al.* (2006) indicate that their calculations are conservative, thus in this scenario Lake Vostok would be an extremely ultra-oligotrophic system with a total organic carbon pool of <250 μmol L^{-1}. Perhaps we need a new term for this level of oligotrophy since the term ultra-oligotrophic is currently applied to extremely unproductive Antarctic surface lakes, that are significantly more productive than the prediction for Lake Vostok.

The second ecological scenario is that life is sustained through a supplemental microbial food web based on chemolithotrophic primary production (Figure 5.9). It has been suggested, based on the presence of a thermophilic bacterium in the accretion ice (see above) and other evidence, that a geothermal system may exist beneath Lake Vostok's waters. Thus a thermophilic chemolithoautotrophic community might exist in faults beneath the lake (Bulat *et al.* 2004). While Christner *et al.* (2006) accept this possibility they argue that a chemolithotrophic-based ecosystem is conceivable without geothermal activity. A range of reduced compounds may be present to support biogeochemical reactions. Oxidants (oxygen and nitrate) would be supplied by the ice sheet and by chemical weathering of bedrock and sediment (e.g. SO_4^{2-} from sulphide oxidation). Basal ice continually melts into the lake, providing glacial debris containing sulphide and iron minerals as well as organic material. This input provides material for biological processes mediated by microorganisms. Such microbially mediated chemical weathering interactions have been described in hypoxic and anoxic glacial beds (Tranter *et al.* 2002a).

In the very near future Lake Vostok and the myriad of other subglacial lakes in Antarctica will yield their secrets. Not only will this tell us more about our planet and microbial ecology, it will also provide insights into how life may be sustained on other planets and their moons.

5.5 Lake Vida

Lake Vida is one of the largest lakes in the McMurdo Dry Valleys of Antarctica and is thought to possess the thickest ice cover of any surface lake worldwide. It is situated in the Victoria Valley and is 3.5 km long and 1km wide. The lake is effectively a small-scale model of a subglacial system. It was thought that Lake Vida was frozen to its base, but an investigation by Doran *et al.* (2003) has revealed that under ~19 m of ice there is a layer of concentrated brine. Using ground penetrating radar (GPR) they were able to produce a bathymetric survey of the lake (Figure 5.10). They were unable to measure the depth of the underlying brine layer because GPR signals are absorbed by saline water, thus the contours on Figure 5.10 do not go below the wet ice layer. However, based on the bathymetry they were able to estimate that the brine layer is around 5 m deep and calculated that it has a temperature of −10°C. Two ice cores were obtained from Lake Vida, one 14m long and a second of 15.8 m. The ice contained bubbles and sediment and microbial mat layers (Figure 5.11). The sediment layers in the top 7 m indicate significant summer flooding, which has clearly been more frequent in the recent history of the lake. Summer flooding from glacial melt cannot flow beneath the ice as it does in many other Dry Valleys lakes where moats develop around the lake edges in summer. In Lake Vida the meltwater contributes to the surface ice. At the same time a net freezing of around 7 cmyear^{-1} may occur at the ice bottom. This differs from the ice covers of other Dry Valleys lakes where the ice covers represent a dynamic equilibrium between downward movement of sediment as a result of melting during the summer, and the upward movement of ice from

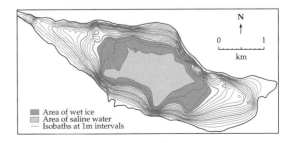

Figure 5.10 Map showing depth contours in Lake Vida and the area of wet ice and saline water. Redrawn from Doran *et al.* (2003).

ablation at the surface and freezing at the bottom of the ice layer (see Section 4.2.1).

The bottom of the 15.8 m core was composed of wet saline ice with a temperature of –11.5°C. The brine trapped in the ice was predominantly NaCl with a salinity of 245‰, the equivalent of ×7 seawater. The ratio of salts was very close to the ratios seen in other Dry Valleys lakes, e.g. Lake Bonney, suggesting that Lake Vida has a similar history. [14]C analyses of microbial mat layers frozen in the ice indicate that the lake's history extends back at least 2800 ^{14}C years. Measurable photosynthesis and bacterial production occurred within water melted from the ice core where organic layers were present (Figure 5.11). The rates were very low, but nonetheless suggest that the microbial communities trapped in the ice for hundreds of years are able to rapidly become metabolically active when liquid water is available.

Low concentrations of chlorophyll *a* were evident in the middle to upper section of the ice core (Table 5.5). Bacterial abundances within the core (Table 5.5) were high compared with other glacial environments (see Table 3.5) and the Greenland Ice

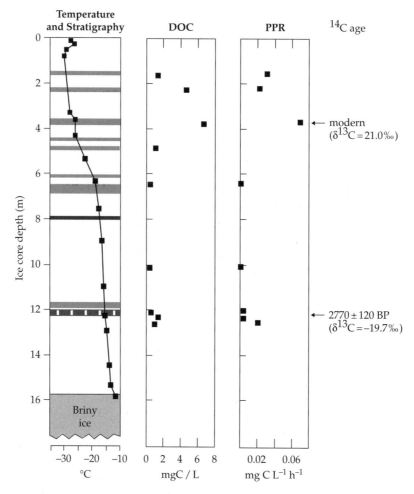

Figure 5.11   Temperature and stratigraphy of the ice cover of Lake Vida (left panel). The dark bands indicate sediment layers, and the band with vertical dashes contains microbial mats. The middle panel shows dissolved organic carbon (DOC) and right hand panel shows primary production rates (PPR) in mg C L^{-1} h^{-1}. Also indicated are ^{14}C dating ages. Redrawn from Doran *et al.* (2003).

Table 5.5 Chlorophyll *a* and bacterial cell abundances in the water melted from a 15.9m ice core taken from the Lake Vida ice cover.

Depth of ice core sample (m)	Chlorophyll *a* (µg L^{-1})	Bacterial cell abundance cells × 10^6 mL^{-1}
4.8	0.3	2.09
6.9	0	1.33
11.8	1.9	0.47
14.7	No data	0.54
15.9	0	0.12

Data from Mosier *et al.* 2007

Core (2.4×10^4m L^{-1}), but within the range reported from sea ice (see Table 4.5). In common with glacial ice (see Section 5.3), most of the bacteria were small with mean cell volumes ranging between 0.028 and 0.083µm^3, the highest value occurring at 15.9 m (Mosier *et al.* 2007).

Molecular analysis of material from sections of the cores (4.8 and 15.9 m) revealed a surprising diversity for what is a very extreme environment, including prokaryotes that are the usual inhabitants of ice (see sections above) and eukaryotic microorganisms. Among the bacteria, the Actinobacteria dominated the clone library at 4.8m (42% of the sequences) with Bacteroidetes contributing 13%, Gammaproteobacteria 10%, and Cyanobacteria 8%. In contrast, in the saline ice at 15.9 m the clone library was dominated by Gammaproteobacteria (52% of clone sequences) with 15% contributed by Bacteroidetes and 11% by Actinobacteria. Around 72% of the sequence derived from the 15.9 m ice was most closely related to phylotypes isolated from polar and glacial habitats including the sea ice.

Interestingly no Archaea were amplified by the archaeal primers used.

Eukaryote denatured gradient gel electrophoresis (DGGE) of samples from a wider range of depths within the core produced sequences dominated by green algae (Chlorophyta) including *Chlorella* spp. and *Chlamydomonas*, as well as Chrysophyceae and Bacillariophyta (diatoms). Ciliated sequences were also detected (Mosier *et al.* 2007). The major question is whether these organisms are metabolically active within the ice. They can become active when the ice is melted, as the photosynthesis and bacterial production data of Doran *et al.* (2003) have demonstrated.

The briny ice in the lower section of the ice profile is likely to harbour halophytic bacteria. Cultures of species related to *Psychrobacter* and *Marinobacter* (Gammaproteobacteria) have been isolated from 15.9 m samples of the ice core (Mondino *et al.* 2009). Cultures allow us to gain an idea of the physiological tolerances and growth characteristics of species. Three of the five isolates were psychrophilic, while the other two were psychrotolerant, but they all had optimum growth temperatures below 20°C (Table 5.6). All of the strains were able to grow at salinities above 10% NaCl and some up to 15% NaCl (Mondino *et al.* 2009). While NaCl is the major salt in the lower Lake Vida ice, other salts are present at significant levels (Mg$_2^+$, K$^+$, and SO$_4^{2-}$). The salinity of the brine in the ice is around 24%, higher than the optimum values in Table 5.4. None of the strains were grown below –4°C, however bacteria can grow at lower temperatures. It highly likely that

Table 5.6 Growth temperatures and salinity tolerances of *Psychrobacter* relatives (strains P1 to P3) and *Marinobacter* relatives (strains M1 and M2).

Strain	Temperatre range (°C)	Temperature optimum (°C)	Salinity range (% NaCl)	Salinity optimum (% NaCl)	Cold/salinity phenotype
P1	–4 to +25	8	2–13	10	Psychrotolerant/halophilic
P2	–8 to +18	0–8*	0–10	5	Psychrophilic/halotolerant
P3	–4 to +15	8	2–10	5–10*	Psychrophilic/halophilic
M1	–4 to +25	14	2–15	10	Psychrotolerant/halophilic
M2	–4 to +20	4.8	2–15	2–5*	Psychrophilic/halophilic

* The mean growth yield was virtually identical at all the temperatures or salinities indicated.
Data from Mondino *et al.* (2009).

these bacteria are metabolically active in the briny ice and in the brine layer underlying the ice. Elsewhere in Antarctica in the coastal Vestfold Hills there are hypersaline lakes with salinities as high as 32%, around ×9 seawater (Deep Lake), and temperatures that drop to –17°C in winter. These lakes support bacterial communities that show seasonal changes in species abundances, suggesting that active growth is occurring even in winter (James *et al.* 1994). Obviously, for growth, an adequate source of carbon is required, which is the case in the Vestfold Hills hypersaline lake systems. As yet there are no published data for DOC in the briny ice of Lake Vida or the brine layer.

CHAPTER 6

Astrobiology

6.1 Introduction

It may, at first glance, seem strange to include a chapter on astrobiology in this volume, but one can barely read an article on life in surface or subglacial environments without encountering a statement that such habitats are an analogue for life on other planets in our solar system, especially Mars. Indeed, a recently published book on *Life in Antarctic Deserts and Other Cold Dry Environments* focuses on these environments as an astrobiological analogue (Doran *et al.* 2010). As we have seen in the preceding chapters, the organisms, mostly microbes, that live on or in ice are adapted to living under extreme conditions and are often referred to as extremophiles. If life exists in our solar system, or indeed elsewhere in the universe, the environments are likely to be extreme and must therefore be colonized by extremophiles.

We tend to view life within the context of what we see on our own planet, but the more modern approach is to define life from a different perspective in astrobiology. The National Aeronautics and Space Administration (NASA) assembled a committee in 1994 to consider possible life in the cosmos and came up with the proposal that life is a 'self-sustaining chemical system capable of Darwinian evolution'. The use of the term 'system' recognizes that entities, for example a virus or cell, can live without individually exemplifying life (Benner 2010). Benner (2010) argued the possibility that in the future we may encounter life in the cosmos that lives under conditions where life on Earth would not survive. He used as an example the manner in which the *Viking* 1976 mission applied life detection tests on Mars. Three tests we are made: first, adding Martian soil to water that contained ^{14}C-labelled organic compounds that would result in the release of radiolabelled CO_2 if life was present. Second, placing Martian soil in nutrient broth, anticipating the release of oxygen if life were present. And third, using radiolabelled CO and CO_2 in the presence of sunlight to determine if carbon fixation from the atmosphere was occurring (as it does during photosynthesis on Earth)—the labelled gases should be fixed into organic material in the soil. All of these tests gave positive results, though perhaps not as conclusive as many would have wanted. Nevertheless they did indicate biological processes. However, the final conclusion was that there was no life because results from gas chromatography and mass spectrometry tests run in tandem on Martian soil did not detect any organic molecules, despite the fact that there was an indication of metabolism from the three 'life tests'. These latter data were attributed to non-biological processes. Benner (2010) made the point that this conclusion says something about the definition/theory of life held by the scientists. The lack of organic material in the soil overrode the metabolic evidence.

In 1996 a paper appeared in *Science* that suggested that relic biogenic activity might be present on a Martian meteorite found in Antarctica (McKay *et al.* 1996). This meteorite arrived on Earth 13000 years ago and was free of terrestrial weathering. It broke relatively easily along pre-existing fractures and it was on these fracture surfaces that polycyclic aromatic hydrocarbons (PAHs) were found. These were similar in texture and size to bacterially induced carbonate precipitates found on Earth. Moreover, high resolution scanning electron microscopy revealed what looked like tiny bacteria about 100nm in diameter. Such nanobacteria have been described as abundant in travertines deposited near hot springs and in other carbonate sediments (Folk 1993). They are also found in Paleozoic and Mesozoic rocks. McKay *et al.* (1996) conceded that there may other explanations for what they observed, but argue that when the data are viewed collectively they provide strong evidence for primitive life on Mars. Some argued that these bacteria were too small to contain ribosomes and construct proteins. The Martian bacteria are much smaller than the ultramicrobacteria described in Section 5.3, which are 0.2–0.3 µm in diameter and occur in subglacial habitats and within glacial ice. However, it is now widely accepted that early life on Earth passed through a stage when RNA both stored genetic information and catalysed chemical reactions, the so-called 'RNA World' (Gilbert 1986). Thus very small microorganisms containing only RNA would be feasible.

For a long time it was assumed that the prebiotic synthesis of the ribonucleotide building blocks of RNA were assembled from their three molecular components: a nucleobase (these being adenine, guanine, cytosine, or uracil), a ribose sugar, and phosphate. However, finding a way to properly join the pyrimidine nucleobases cytosine and uracil to ribose proved challenging. It was almost impossible to envisage the spontaneous assembly of a complex RNA molecule, which led to a search for simpler genetic polymers that may have preceded RNA during the evolution of life (Szostak 2009). In 2009, Powner *et al.* showed that activated pyrimidine ribonucleotides can be formed in short sequences that bypass free ribose and the nucleobases, instead proceeding through arabinose amino-oxazoline and anhydronucleoside intermediates. The materials for the synthesis were all plausible feedstock molecules under conditions that fit early Earth geochemical models. During evolution, RNA was the intermediate between DNA and proteins. That life could have originated from the abiotic emergence of RNA has now been demonstrated. Thus when considering life on other planets we need to hold a broader perspective of life. Nonetheless, useful analogies between conditions found in extreme ice and aquatic environments on Earth, and the life forms supported within them, can be drawn with potential life-supporting environments on other planets.

Cockell *et al.* (2011) conclude that one of the most remarkable discoveries derived from the exploration of space is the likely presence of liquid water

under ice deposits on other planetary bodies—extraterrestrial subglacial environments. These include the ice-covered oceans of the Jovian moon, Europa, and the Saturnian moon, Enceladus. There is no current evidence for subglacial liquid water on Mars today, but more favourable conditions may have existed in the past. These existing and past extraterrestrial subglacial environments share some similarities with subglacial environments on the Earth, but they also have many marked differences. Cockell *et al.* (2011) suggest that extraterrestrial environments may provide three new types of sub-glacial settings for study: (i) uninhabitable environments that are more extreme and life limiting than terrestrial subglacial environments; (ii) environments that are habitable, but are uninhabited, which can be compared to similar biotically influenced subglacial environments on Earth; and (iii) environments with examples of life that will provide new opportunities to investigate the interactions between biota and glacial environments. Finally, Tranter *et al.* (2010) suggested that supraglacial environments on the Martian polar ice caps are currently uninhabitable, but that cryoconite hole-like environments were possibly habitable should surface energy balances have been more positive in the past. We summarize the main findings from these papers below.

6.2 Extraterrestrial cryospheric environments

6.2.1 Mars

Currently, Mars has ice caps at both its North and South Poles. Enough water is locked up in the South Pole to cover the planet in a liquid layer ~11m deep (Plaut *et al.* 2007). There is significant annual growth and retreat of the polar caps, which expand to cover ~30% of the planet's surface during the winter and shrink to relatively small caps covering ~1% of the surface in summer. The perennial portion of the North Pole cap consists almost entirely of water ice (Langevin *et al.* 2005). This gains a seasonal coating of frozen CO_2, which grows to ~1m thick in winter. The South Pole cap consists of two layers. The top layer consists of frozen CO_2 and is ~8m thick. The bottom layer is very much deeper and is made of water ice (Titus *et al.* 2003). Flat-floored, circular pits

~8m deep and 200–1000m in diameter occur within the upper frozen CO_2 layer during the summer, revealing the water ice below (Byrne and Ingersoll 2003). Thus, although the two polar caps are similar, the significant difference between them is that the South Pole dry ice cover is thicker and does not completely disappear during the summertime.

The geomorphology of the Martian surface suggests that the climate was warmer and wetter in the past. Pertinent features include ancient valley networks (e.g. Fanale *et al.* 1992; Mangold *et al.* 2004), outflow channels (Baker 1982), and deltas (Pondrelli *et al.* 2008). Some believe that Mars might have once possessed an ocean (Parker *et al.* 1989; Taylor Perron *et al.* 2007), occupying the low-standing northern hemisphere, although this is still contentious (e.g. Carr and Head 2003).

At present, the upper few kilometres of the Martian crust contain large amounts of water ice (Squyres *et al.* 1992), and the surface regolith can contain more than 50% ice by volume (Boynton *et al.* 2002; Feldman *et al.* 2004), covered by a few centimetres of ice-rich dust (Smith *et al.* 2009). The ice persists to depths of a few kilometres, at which point geothermal temperatures are high enough for liquid water to be present. The top of the ice table is driven deeper at lower latitudes, but the base of the ice table is shallower, because of warmer year-round surface temperatures (Fanale 1976). Cockell *et al.* (2011) identify the water at the base of the ice table as the Martian cryospheric environment most conducive for life, since the water could remain as a liquid for geologically significant time periods. A second possible acquatice location is at the margins of the North Pole cap, although this needs a substantial amount of solute to allow melting of the kilometre-thick, perennial, polar water ice cap (Phillips *et al.* 2008), since modelling suggests that pressure melting is unlikely otherwise (Greve *et al.* 2004). To date, radar studies of massive ice deposits on Mars have not revealed liquid water (e.g. Holt *et al.* 2008). However, Cockell *et al.* (2011) caution that thin layers of water at the base of glaciers would not be easily visible to radar analysis. Other smaller Martian glaciers are inferred to be cold based with no basal melting (Head and Marchant 2003; Shean *et al.* 2005), and it is thought that warm- or wet-based glaciers were present on Mars around a bil-

lion years ago, possibly on the highlands surrounding the Argyre and Hellas impact basins (e.g. Kargel and Strom 1991; Banks *et al.* 2008). Finally, there might have been an equatorial 'frozen sea' in the Elysium Planitia region of Mars a few million years ago (Burr *et al.* 2002b; Murray *et al.* 2005), arising from the slow freezing of liquids emanating from a deep tectonic fracture (Burr *et al.* 2002a; Head *et al.* 2003).

Cockell *et al.* (2011) speculate that the wet base to the ice table, described above, may contain suitable REDOX pairs and nutrients to sustain a microbial ecosystem. The environment is likely to be anoxic, given the low oxygen content of the Martian atmosphere. A range of electron acceptors also found in terrestrial subglacial environments are likely to be found in Martian regolith, including Fe(III) and Mn(IV), both from comminuted basaltic rocks. There are also plentiful supplies of sulphate (Clark *et al.* 1982; Rieder *et al.* 1997; Gendrin *et al.* 2005; Langevin *et al.* 2005; Squyres *et al.* 2006). Electron donors could include Fe(II), produced from comminuted basalts, and organic material delivered exogenously in meteorites. Bioessential trace elements, such as zinc, copper, and nickel, are found in glacially comminuted basaltic rocks, and phosphorus can be found in apatite. Sources of nitrogen are more problematic. However, fixed nitrogen could have been produced by volcanic or impact processing in the early history of the planet (Segura and Navarro-Gonzalez 2005; Summers and Khare 2007; Manning *et al.* 2009), with nitrate-containing minerals subsequently made available by glacial comminution.

Cockell *et al.* (2011) conclude that Martian subglacial environments would be favourable places for life whenever melting occurs, analogous to microbial activity in present-day subglacial environments on Earth. Wider scale melting was more likely in the past than at present. However, they believe that the search for extant life in Martian subglacial environments is a valid objective, and that subglacial environments are also favourable locations to search for past life on Mars.

The possibility of supraglacial habitats, specifically cryoconite holes, in the polar ice caps was explored by Tranter *et al.* (2010). Reddish atmospheric dust, with a mean diameter of <5 µm, is ubiquitous in the

Martian atmosphere (Goetz *et al.* 2005). Dust deposited onto and contained within the ice caps gives them a reddish colour (Langevin *et al.* 2005). Another source of dust to the ice cap surfaces is that locally redistributed by the CO_2 jets (Figure 6.1) that emerge from the Southern Ice Cap each spring (Kieffer *et al.* 2006) and which spray fine dark sand hundreds of feet up into the air and around their vent holes. These jets are believed to form the dark spots or fan-like features that form on the Southern Ice Cap surface each spring, and which gradually evolve into spider-like features as the summer progresses (Figure 6.2). The spots are typically 15–50 m wide and a 100 m or so apart and usually form in similar locations on an annual basis. The combination of dust-laden water ice would superficially suggest that there is the potential for water production by the heating of subsurface dust within the polar ice caps during the summer. The overlying ice would protect the water from freezing or sublimation into the cold, dry atmosphere, in a similar manner to the protection offered to water produced in the subsurface drainage system of the Dry Valleys glaciers.

Unfortunately, the energy balance conditions in the ice caps are not conducive to the production of liquid water (MEPAG 2006). Calculations suggest that the maximum possible temperature that subsurface water ice could reach is –43°C, even with favourable aspect and latitude. Hence, present-day ice cap surface environments are deemed to be solidly frozen and incapable of producing liquid water as a consequence of the solar heating of subsurface debris.

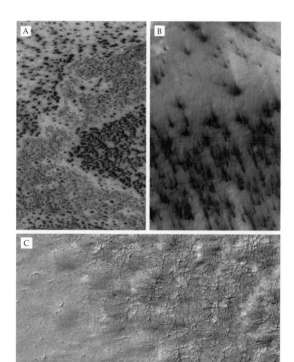

Figure 6.2 Dark spots (A) and fans (B) on the surface of the Polar Ice Cap on Mars. (C) Spiders form on the top of the residual polar cap, after the seasonal CO_2 ice slab has disappeared. Next spring, these will probably mark the sites of vents when the CO_2 ice cap returns. Each image is ~3 km wide. Image courtesy of NASA/JPL/MSSS.

Great excitement followed the discovery of methane in the Martian atmosphere (Formisano *et al.* 2004). We do not yet know if the methane on Mars is biogenic in origin. On Earth the isotopic signatures

Figure 6.1 Sand-laden CO_2 jets that are believed to emanate from the Southern Ice Cap of Mars during the spring. Image courtesy of Ron Miller, Arizona State University.

CHAPTER 7

Future directions

7.1 Introduction

The purpose of this chapter is to reveal where future research will be directed and to indicate which techniques and approaches are likely to be implemented. Where possible, integrative research will also be outlined, because the rapid development of cryospheric ecology means that our understanding is presently fragmented. This situation is typical of fast-moving scientific enquiry and arises from the lack of interdisciplinary research interaction and a narrow base of practitioners. More interdisciplinarity is particularly important for scientific development because tight interactions between physical, chemical, and biological processes are key features of cryospheric ecosystems (e.g. Fountain and Tranter 2008). There are exceptions however: strong interactions between microbial ecologists, biogeochemists, and geoscientists are a key feature of the early stages of subglacial Antarctic lake exploration (Siegert *et al.* 2006). These interactions have been occurring for some time, but largely in places where major research infrastructure provides the necessary field and laboratory support. For example, research stations established by New Zealand and the USA in the McMurdo area, and the National Science Foundation funded Long Term Ecosystem Research Program (LTER), have contributed greatly to the investigation of perennially ice-covered lakes and ice shelf ecosystems (Howard-Williams *et al.* 1990; Priscu *et al.* 1998). This logistical advantage is also significant in the case of marine cryospheric ecosystems, which make use of research vessels in the open sea and small boats deployed from bases to the coastal environment. Significant examples of the latter are most commonplace in the maritime Antarctic, such as the Admiralty Bay ecosystem of

King George Island (Clarke and Leakey 1996; Corbisier *et al.* 2004), where multinational collaboration at a single field site is taking place. However, very important coastal environments in West Antarctica and other parts of the polar regions remain logistically challenging and therefore poorly understood. For example, some of the greatest terrestrial inputs of organic carbon and other nutrients such as iron into coastal Antarctic ecosystems most likely occur in the Pine Island and Thwaites Glacier areas of West Antarctica, where recent ice mass losses have been in excess of 200 Gt year^{-1} (Rignot *et al.* 2008). Accessing this near-coastal environment is sometimes impossible for research vessels and is more likely to be achieved opportunistically, rather than as part of a low-risk voyage plan. Improving the geographic distribution of research to include areas like this is an important priority for the next decades of cryospheric ecology, especially in Antarctica. It requires both an increase in the remote sensing of isolated habitats and more ambitious field monitoring studies to be undertaken.

In several cases, the use of remote sensing is already maturing as a tool for understanding ice-bound ecosystems, whilst detailed ground studies are conspicuous by their absence. For example, the distribution of wet snow across the Greenland and Antarctic Ice Sheets is now reasonably well understood from passive microwave observations (e.g. Waleed and Steffen 1997), but *in situ* measurements of ecosystem functioning are only just beginning. In other areas where remote sensing from satellite platforms are less likely to succeed there exists a need to resort to geophysical surveys, modelling, and technological developments to enable *in situ* exploration. For example, unseen habitats such as wet sedimentary environments at the bed of the

Antarctic and Greenland Ice Sheets remain virtually unexplored, yet recent studies suggest they are likely to harbour far more life than subglacial lakes (Lanoil *et al.* 2009). Better insights into the activity of microorganisms must also be sought using improved *in situ* measurements, especially in the case of glacial habitats. Interest in subglacial Antarctic lakes means that protocols for sampling these habitats are already in place (e.g. Christner *et al.* 2005b; Doran *et al.* 2008), but *in situ* measurements of microbial activity remain problematic, especially in the case of ice sheets. This necessitates the development of new sensor technology, including sensor networks for deployment down crevasses and boreholes and optical methods for life detection along ice cores, across ice surfaces, and within lakes (Price 2000; Sattler *et al.* 2010). Technological advances can also be expected to enhance the study of other icy habitats, especially sea ice, where habitat heterogeneity and logistical difficulties need to be overcome before *in situ* observations are common-place (Mock and Thomas 2005). At the time of writing, there is great emphasis being placed upon the deployment of sensors using remote controlled and autonomous platforms. These will greatly improve data acquisition from environments that are either too inaccessible or risky for people to access. Integration of these technological advances into cryospheric ecology is another reason why the interdisciplinary approach is key—this time bringing engineering expertise into the forum. Presently, much of the energy behind the integration of the latter expertise results from the importance of cryospheric ecosystems as analogues for other icy planets (Storrie-Lombardi *et al.* 2008).

Models characterizing rock–water–microbe interactions and nutrient cycles through cryospheric ecosystems also need development, so that better use can be made of existing data sets. This is particularly so in the context of glacial ecosystems, which are far less understood than ecosystems in other habitats, yet sufficiently data-rich to allow early model development (e.g. Roberts *et al.* 2010). They are also arguably more amenable to modelling than sea ice ecosystems, yet here there are sophisticated ecological models already in use (e.g. Pogson *et al.* 2011). Models intended for all cryospheric habitats also need to benefit from the rapid develop-

ment of modern molecular methods for the analysis of biological material. In so doing, cryospheric ecologists will be able to move on from the 'who's there' insights gained from DNA sequencing, and begin properly integrating nutrient mass balance data sets with molecular information that describes microbial function and activity. These methods will also present important opportunities for understanding the role cryospheric habitats have played in the evolution of these microorganisms, not least because the residence time of cells within polar ice sheets can be >10^6 years (Priscu *et al.* 2006). This time scale is approaching that required for the survival of microorganisms through Snowball Earth events, which are also intricately linked to the evolution of microorganisms upon early Earth (Vincent and Howard-Williams 2000).

7.2 Priority field sites for future research

Logistically easier conditions in the Arctic mean that cryospheric ecology is almost as well established here as it is in alpine studies conducted at lower latitudes. However, there are exceptions, since the subglacial ecosystems of the Greenland Ice Sheet are almost less understood than those of Antarctica, where subglacial lake science is now more than a decade old. Instead, insights into subglacial (and surface-derived) microbial consortia have been derived from the analysis of Greenland Ice Sheet ice cores (e.g. Tung *et al.* 2006; Miteva *et al.* 2009). However, these ice cores have not been collected for ecological purposes, and so cannot yet be used to establish the biogeography of these ice sheets until their locational bias in favour of accumulation areas has been assessed (Figure 7.1). This under-sampling problem can be addressed by ice coring at several points along glacial flow lines in the major ice drainage basins of the Antarctic and Greenland Ice Sheets (Figure 7.1). Variations in the microbial consortia at different depths of the ice core can then be combined from several locations along the flow lines to establish the influence of changing climate and location upon the biogeography of ice sheet ecosystems. So far, only variations in depth have been assessed, showing that changes in the microorganisms entrapped during different

A

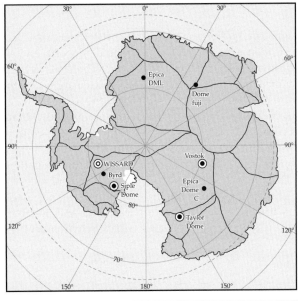

B

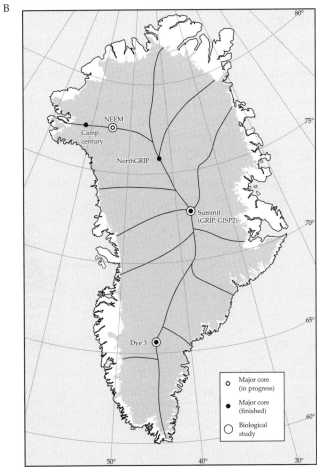

Figure 7.1 Deep ice core drilling sites upon the Antarctic (A) and Greenland (B) Ice Sheets, showing which cores have been subject to biological study. The boundaries of the major ice drainage basins are also given; note how Greenland cores are located on or near drainage divides. DML, Drønning Maud Land; GISP, Greenland Ice Sheet Project; GRIP, Greenland Ice Core Project; NEEM, North Greenland Eemian Ice Drilling; WISSARD, Whillans Ice Stream Subglacial Access Research Drilling.

climate periods vary significantly. For example, Miteva *et al.* (2009) found evidence for a significant increase in total cell deposition during colder climates due to enhanced dust transport over the Greenland Ice Sheet (Figure 7.2). The importance of dust is well known from the study of glacier surface ecosystems (see Sections 3.1.1 and 3.2.3) and has also been revealed by the fluorescence study of several sections of Antarctic and Greenland Ice Cores (Price *et al.* 2009). The inference, therefore, is that terrestrial inoculi of ice sheet ecosystems might be more important than marine inoculi. Local, terrestrial inoculi are most likely to be dominant near ice sheet margins (e.g. Stibal *et al.* 2010) and so it would be worth establishing whether more exotic microorganisms are apparent at great distances from ice margins. Further, it is unclear what biogeography we might we expect where ice sheets have marine margins. Microbiological characterization of more ice cores collected along ice sheet flow lines would therefore help capture hitherto unknown biogeographical patterns. It would also have the added value of improving our quantitative understanding of how glaciers and ice sheets deliver nutrients and microorganisms directly into the marine environment.

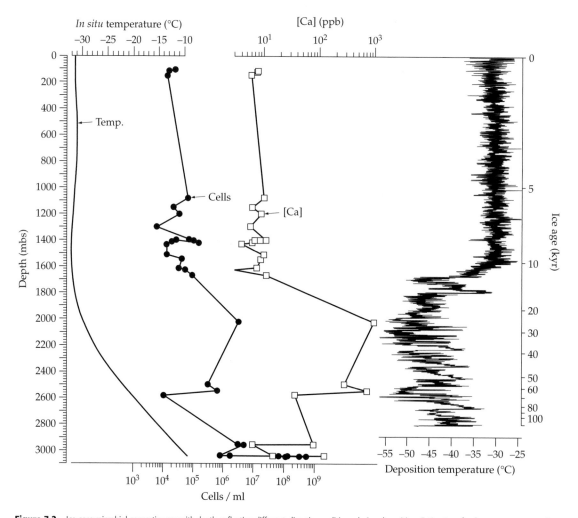

Figure 7.2 Ice core microbial consortia vary with depth, reflecting different climatic conditions during deposition. Estimates of palaeotemperature and dust deposition (the latter represented by Ca²⁺) enabled Miteva *et al.* (2009) to establish the links between climate change and microbial accumulation in the GISP2 ice core. mbs, metres below surface. Redrawn from Miteva *et al.* (2009).

A better characterization of sub-ice sheet ecosystems also requires the systematic study of basal ice exposures along the periphery of ice sheets and a concerted effort to retrieve more subglacial sediment cores. Presently, the geomicrobiology and biogeochemistry of basal ice are best developed in the case of outlet glaciers from the Greenland and Antarctic Ice Sheets, where ice exposures may be cleaned and readily sampled (e.g. Yde *et al.* 2010). Sediment cores from subglacial environments are also a priority, because only one core has been analysed opportunistically and described in the scientific literature at the time of writing (Lanoil *et al.* 2009). At present there are several major drilling initiatives planned for Antarctica, a significant example being the Whillans Ice Stream Subglacial Access Research Drilling (WISSARD) project scheduled for completion in 2015 (http://www.wissard.org/). Other sediment cores are also likely to be recovered when drilling to subglacial lakes has been successful.

The requirement for a greater spatial distribution of sampling sites is also relevant in the context of the other key cryospheric habitats, most notably snow and ice shelves. For example, while the ecology of snow is well established in terms of its relationship with aquatic and terrestrial ecosystems in temperate and boreal environments (Jones 1999), there are virtually no studies of snow pack ecosystem production upon ice sheets. Ice shelf studies are heavily biased toward the ecosystems upon the Ross Ice Shelf in Antarctica, and the Ward Hunt and Markham Ice Shelves in the Canadian Arctic. Several major ecosystems are therefore unexplored, including the Amery, Larsen, King George VI, Filchner, and Ronne Ice Shelves. These environments are likely to make excellent research sites. For example, the Amery Ice Shelf contains the largest blue ice area on Earth (Liston and Winther 2005), so active, photosynthesizing, microbial communities similar to those found on blue ice of the nearby Vestfold Hills can be expected (see Section 3.2.4.2). Upon the Larsen Ice Shelves, the accumulation of water is very pronounced and most likely contributed to the disintegration of Larsen A and B into the Weddell Sea in 1995 and 2002, respectively (e.g. Scambos *et al.* 2003). Elsewhere in maritime Antarctic, high rates of regional climate warming have greatly increased the

amount of surface melting, suggesting that increased biological production upon ice shelves requires continued research attention. Upon the King George VI Ice Shelf, dust deposition events are known to induce melting during summer (Wager 1972), resulting in water accumulation and providing a nutrient source for ecosystem activity at the same time. This dust-fuelled ecosystem offers a potential analogue for studying life during the hypothesized Snowball Earth conditions (cf. Vincent *et al.* 2000) and so its study is easily justified.

7.3 Remote sensing development

In the future, remote sensing from satellite and airborne platforms will continue to provide vital details about the distribution of cryospheric habitats. Recent, major advances in this context have included the European Space Agency's new CryoSat II mission, which will greatly enhance the provision of elevation and thickness data for polar ice. These data are critical for quantifying the mass of ice in cryospheric habitats. Further, changes in altimetry are now being used to document hydrological coupling between subglacial habitats (e.g. Wingham *et al.* 2006), and will be increasingly used in combination with airborne radar to relate this coupling to the distribution of wet sedimentary environments that are likely to harbour large microbial populations (Fricker *et al.* 2010). Use of the National Aeronautics and Space Administration (NASA) ICESat platform is already achieving this by establishing the geography of dynamic (rather than static) subglacial lakes beneath the Antarctic Ice Sheet (Smith *et al.* 2009).

The use of more direct biological remote sensing methods remains largely confined to the ocean–sea ice system. For example, chlorophyll mapping can be achieved from satellites using products such as SeaWifs (http://oceancolor.gsfc.nasa.gov/SeaWiFS) and has recently been used to reveal evidence for ocean fertilization in the wake of icebergs (Schwarz and Schodlok 2009). SeaWifs has also been used to establish the importance of coastal polynyas as hotspots for biological production (Arrigo and van Dijken 2003). However, there are virtually no applications of this type of analysis to snow and ice surface ecosystems. Two studies show great potential with respect to the remote sensing of large snow algal blooms: an airborne study by Painter *et al.* (2001) and a satellite remote sensing study by Takeuchi *et al.* (2006). In the latter study, the spatial distribution of algae across the Harding Icefield was quantified using a reflectance ratio of SPOT (Satellite Probatoire d'Observation de la Terre) satellite band wavelength 610–680 nm to band 500–590 nm. Calibration was shown to be possible using a field spectrometer and snow samples that were enumerated using microscopy. These studies suggest that long-term monitoring of intense snow algal blooms can reveal significant, new insights into the close relationship between climate and primary production on the surface of snow cover. However, its application remains to be better exploited, especially within maritime Antarctica, where intense algal blooms occur on coastal snow packs. Here, however, cloud cover might compromise data capture. Certain ice shelves in both polar regions are also known for their high biomass cover (e.g. Vincent *et al.* 2004), offering further potential for remote sensing studies. However, before these techniques can be employed successfully over ice sheets and glaciers, their detection limits and spatial resolution need to be improved, because the chlorophyll cover is both lower and less continuous. The use of airborne, near-surface platforms is therefore preferable to orbiting satellites if these data requirements are to be met. Remote sensing from remote controlled and autonomous aircraft offers considerable potential (Figure 7.3). Such intermediate platforms also provide an important means of quantifying cryospheric ecosystems in environments too hazardous or inaccessible to researchers. Both unmanned airborne and underwater vehicles (UAVs and UUVs, respectively) will be increasingly deployed with on-board life detection sensors. Astrobiological and oceanographic sciences are presently providing the enabling technologies for these applications. For example, several projects describe tethered and autonomous platforms for exploring subglacial lakes, once they have been penetrated (e.g. Siegert *et al.* 2006). These platforms offer the potential for sampling, sensing, data capture, and transmission to the operator. Perhaps the most ambitious example is the 'Cryobot' (Figure 7.4), an autonomous robot that is designed

Figure 7.3 The British Antarctic Survey autonomous aerial vehicle in flight over Antarctica. Image courtesy of British Antarctic Survey.

to melt its own way down into the ice covers of other planets such as Europa (Zimmerman and Feldman 2002).

Hodson *et al.* (2007) used a remote controlled helicopter equipped with a digital camera to demonstrate the advantage of UAVs for data capture, by producing maps of the mass distribution of cryoconite over the Arctic glacier Midtre Lovénbreen. The helicopter was ideal because the stall speed of many fixed wing, remote control aircraft is rather too high for capturing high resolution images of deep cryoconite holes. The images also allowed the habitat form of the cryoconite to be captured, revealing that cryoconite holes were dominant in the upper part of the glacier, whilst stream deposits were dominant in the higher melt zone of the ablation area. Recent transects across the melt zone of the Greenland Ice Sheet described by Bøggild *et al.* (2010), Hodson *et al.* (2010a), and Stibal *et al.* (2010) clearly suggest that UAVs need to be employed for mapping cryoconite at the ice sheet scale. Better characterization of this distribution will also be to the benefit of glaciologists seeking improved algorithms for estimating ice albedo, especially since the end of summer snow lines will move up-glacier with continued climate warming (Bøggild *et al.* 2010).

The future role of UAV and UUV platforms for examining glacial cryospheric ecosystems is likely to be important, yet at the time of writing the number of research papers that describe their potential is still way in excess of the number of papers that report their successful deployment. Further, the use of such platforms for the collection of geospatial and biological

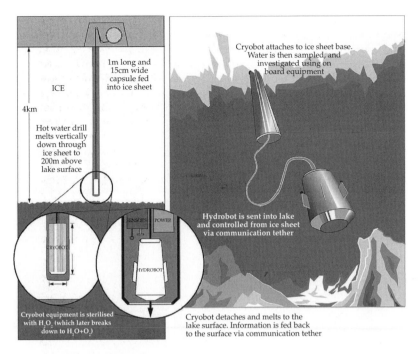

Figure 7.4 The Cryobot concept for subglacial lake exploration. Image courtesy of M.J. Siegert.

data has developed furthest in the context of submarine rather than airborne research. A more comprehensive review of these limnological and marine applications using UUVs is given by Laybourn-Parry and Vincent (2008).

7.4 Sensor technology

In situ quantification of biological and biogeochemical processes within cryospheric ecosystems will be a strong research driver in the future, placing great emphasis upon improved sensor technology. Of particular relevance are the optical sensors, which serve two key purposes: organic matter characterization using fluorescence emission and reagent-free assays of aqueous nutrient and dissolved gas concentrations.

Ultraviolet (UV) induced fluorescence is very likely to become a standard method on account of its ease of integration into UAV and UUV platforms (Fantoni *et al.* 2004). The technique has already been used to map microorganisms entombed in the perennial ice cover of Antarctic lakes (Storrie-Lombardi and Sattler 2009; Sattler *et al.* 2010). Since it is both

non-invasive and non-destructive, it offers the advantage of being able to track biomass change through time and to examine three-dimensional assemblages of microorganisms within the ice matrix. Its future application will therefore allow a better appreciation of how thermodynamic processes influence the location and efficacy of biological processes within icy matrices. The basis of laser-induced fluorescence emission (LIFE) is well described by Sattler *et al.* (2010). One example of this method involves laser excitation between 220 and 250nm, producing fluorescence emission from nucleic acids and aromatic amino acids that can be detected by a UV-sensitive camera (Nealson *et al.* 2002; Bhartia *et al.* 2008; Storrie-Lombardi *et al.* 2009). The method is easily executed, even making use of a pen laser and a commercially mass-produced digital camera for sensing (e.g. Storrie-Lombardi and Sattler 2009) (Figure 7.5). The Sattler *et al.* (2010) study also demonstrated the potential for direct quantification of the biomass—or at least the photo pigments present within it. However, before these methods become reliable, it is likely that problems associated with the target's

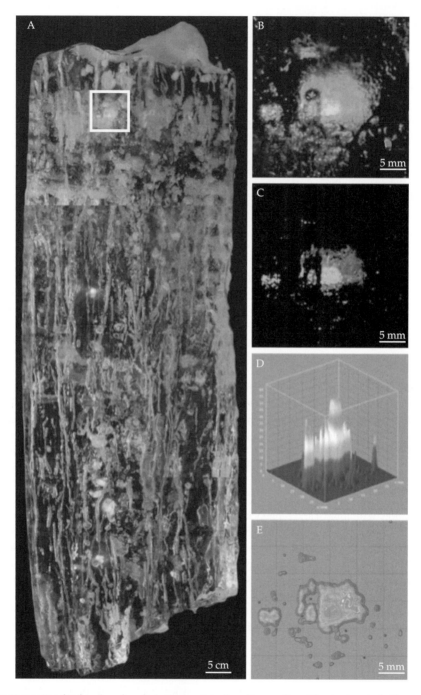

Figure 7.5 (A) A 65 cm section of ice from the surface of Lake Untersee, Antarctica, with cryoconite particles at various depths. Fluorescence was stimulated using a 532 nm laser. (B) The reflected laser light from the target at the top of the core in (A) (see box near top of image), and its fluorescence (C). (D, E) The processed images in 'light cone' and isobar formats, respectively. From Storrie-Lombardi and Sattler (2009), with permission of Mary Ann Liebert Inc. See also Plate 15.

Functional metagenomics focuses on identifying genes that express a function and requires a faithful transcription and translation of the gene or genes of interest and secretion of the gene product, if the screening or assay is to be extracellular. However, the frequency of metagenomic clones that express a given activity is low (Handelsman 2004). In extremophile organisms, functional genomics has identified novel biochemicals such as degradation enzymes (Healy *et al.* 1995) and cold-adaptation proteins (Saunders *et al.* 2003).

Proteomics is another powerful tool in understanding the function dynamics of organisms. This is the large-scale study of proteins, specifically their structure and function. Proteomics has grown at a rapid rate. It relies on mass spectrometry to identify proteins using PMF (peptide mass fingerprinting) or LC-MS/MS (liquid chromatography mass spectrometry) used in combination with stable isotope labeling and two-dimensional gel-based approaches combined with different staining or traditional visualization techniques. It has been argued that the rapid growth of this area has led to a lack of stringency in proteomic data generation and analysis, leading to the publication of some questionable data, requiring the laying down of clear guidelines (Wilkins *et al.* 2006). An example of proteomics elucidating how extremophiles adapt to a range of temperatures is that of the extremophile *Methanococcoides burtonii*, an Archaea. This species grows rapidly at 23°C and can survive in temperatures up to 29°C. The ability to grow faster at temperatures well outside their environmental range is not uncommon among polar prokaryotes. While *M. burtonii* will grow at higher temperatures, it is stressed. The cold adaptation protein in *M. burtonii* has been identified by Goodchild *et al.* (2004); they identified proteins specific to growth at 4°C versus growth at 23°C. Crucial features of cold adaptation involved transcription, protein folding, and metabolism, including the specific activity of RNA polymerase subunit E, a response regulator and peptidyl-prolyl *cis/trans* isomerase. A heat shock protein was expressed during growth at 24°C, indicating stress.

Complete genomes of organisms are being published all the time. The first complete bacterial genome was published in 1995 for the pathogenic bacterium *Haemophilus influenza* (Fleischmann *et al.* 1995). At the time of writing there are over a thousand complete genome sequences available. This has led to the development of comparative genomics, allowing generalizations on genomic organization and evolution to be developed. A particularly important issue that has become apparent from comparative genomics is the phenomenon of horizontal gene transfer (see Section 7.7). One of the major problems is that genome sequencing requires a clonal culture, and only a small proportion of bacteria can be cultured. Sequencing a genome of an uncultured organism is very difficult and has only been achieved on a few occasions (Koonin and Wolf 2008). Bacterial and archaeal genomes show common principles. The sequenced bacterial genomes span two orders of magnitude in size from 180 kb to around 13 Mb. There is a clear-cut bimodal distribution of genome size with the highest peak at ~2 Mb and a second smaller peak at ~5 Mb, suggesting two classes of bacteria: a class with large genomes and a class with small genomes. In contrast the Archaea are less diverse in genome size, ranging from ~0.5 to ~5 Mb, with one peak at around 2 Mb (Koonin and Wolf 2008).

The whole area of metagenomics and proteomics has advanced at an extraordinary pace in the last few decades. Technology is moving forward all the time, allowing greater and more rapid analyses. The study of microbial communities in extreme cold environments now needs to focus much more on how these organisms function and their contribution to carbon and nutrient cycling.

7.7 Elucidating the evolution of extremophile communities

Ribosomal RNA genes are ubiquitous and there is now an enormous database of rRNA sequences. Carl Woese (1987) showed that rRNA genes provide an evolutionary chronometer or clock, which led to the development of the RNA tree for determining microbial relationships. However, the growing genome database for Bacteria has not led to better resolution, but rather to conflicting views. It has been argued that some of the basic tenets of modern biology, that of the Darwinian–Mendelian model of parent to offspring gene flow (vertical gene flow), have been radically challenged for microbes (Charlebois *et al.* 2003). The transfer of genes between organisms other than by vertical gene transfer was recognized before the genome revolution. However, it was not until many microbial genomes were sequenced that the importance of the phenomenon of horizontal gene transfer (HGT) became fully appreciated. Indeed it is clear that it is a dominant force in prokaryote evolution and probably also occurs in eukaryotic microorganisms.

HGT is the transfer of genes between unrelated organisms. This is mediated by so-called mobilomes, which are viruses, plasmids, and other elements that are in constant exchange with more stable chromosomes and act as HGT vehicles. Comparative genomics has revealed that HGT occurs not only between closely related organisms but also between distantly related organisms (Koonin and Wolf 2008). There are many examples, but perhaps one of the most striking is the photosynthetic gene clusters of the Cyanobacteria and other photosynthetic bacteria. There are five phyla of bacteria that have photosynthetic members; these include the Cyanobacteria, the Proteobacteria (purple bacteria), green sulphur bacteria, green filamentous bacteria, and Gram-positive Heliobacteria. It has been proposed that HGT has been the major route for the evolution of bacterial phototrophs. Comparison of the genomes of phototrophs from the different phyla has shown that the genetic components of the photosynthetic apparatus have crossed species lines in a non-vertical manner (Raymond *et al.* 2002). There are also examples of HGT between Bacteria and Archaea.

Where microbes occupy the same habitat there is ample opportunity for HGT to occur. We have considerable evidence of bacteriophages in extremophile prokaryote communities where high rates of infection occur (Säwström *et al.* 2008). Bacteriophages will undoubtedly play an important role in HGT in these environments. As indicated in preceding chapters (e.g. Section 4.1.2), there are a range of important adaptations essential for survival in cold environments such as cold shock proteins, antifreeze proteins, and high levels of polyunsaturated fatty acids that maintain the fluidity of the cell membranes. Comparative genomics has the potential to provide us with a picture of how life in extreme environments evolved in different bacterial, archaeal, and protozoan groups. As Koonin and Wolf (2008) have pointed out, HGT is ubiquitous in the prokaryote world allowing gene transfer between unrelated organisms. The prokaryote world is essentially a single, connected gene pool, although this pool has a complex, compartmentalized structure, with its distinct parts being partially isolated from each other.

Glossary

Ablation—the loss of snow or ice after melting, sublimation, wind erosion, or iceberg production (calving) has taken place. The part of the glacier where these processes exceed accumulation is referred to as the ablation area or ablation zone.

Accumulation—the accumulation of snow and ice on a glacier following snowfall, snow redistribution by wind, and the refreezing of liquid water.

Albedo—the proportion of incident radiation that is reflected by a surface layer.

Archaea—one of three domains in the living world (the others being the Bacteria and Eukaryotes). These unicellular prokaryotes inhabit some of the most extreme environments on the planet, as well as non-extreme habitats.

Assimilation number—chlorophyll a-specific photosynthetic rate defined as the units of carbon fixed per unit of chlorophyll a (e.g. µg C fixed (µg Chl a)$^{-1}$ h^{-1}).

Autotroph—a photosynthetic or chemosynthetic organism.

Bacteria—single-celled organisms known as prokaryotes where the nuclear material is not contained in a nucleus. Within the three domain classification of the living world, Bacteria is one of the domains (the other two being the Archaea and Eukaryotes).

Bacterial production—refers to the growth of a bacterial community and is usually given in units of carbon per unit time per volume of water or sediment, measured by the incorporation of radiolabelled thymidine into DNA or radiochemical-labelled leucine into protein.

Basal freezing—a term that refers to the refreezing of basal meltwater or subglacial lake water. The freezing often results in distinctive basal ice layers that are known for their different ice structure, debris content, and biogeo-chemistry.

Basal melting—ablation from the bed of a glacier, ice sheet, or ice shelf. In the case of terrestrial ice masses, melting occurs due to friction and geothermal heating and is typically less than 2 cm per year. In the case of ice shelves, far greater rates of basal melting, metres per year, are possible when the ice begins to float on top of warmer ocean waters.

Biofilms—thin layers of material containing bacteria, protozoa, and fungi in a matrix of 'biological glue' exuded by the microorganisms. Biofilms form where there is liquid water.

Biogeochemical cycling—biologically mediated transfer of elements, such as carbon, nitrogen, phosphorus, and iron, between inorganic and organic phases, which may be both dissolved and/or particulate. Biogeochemical cycling in the cryosphere is usually microbially mediated.

Blue ice—glacier ice absorbs red wavelengths of light about six times more effectively than blue light, and so the wavelengths of light that are best transmitted through it give ice its characteristic blue colour. This is most apparent when dense (old) ice crystals are exposed by either fracturing or wind polishing, the latter causing Antarctica's well-known blue ice areas. The snow that surrounds these areas appears white because the air between the small crystals scatters and reflects nearly all of the sunlight incident upon them.

Calyptopsis—the second stage in the larval development of euphausiid crustaceans (krill), which metamorphoses into the furcilia stage.

Carotenoids—a pigment, found in cyanobacteria and other photosynthetic organisms like the snow algae, that provides protection against high solar irradiance levels.

Chemotrophs—Bacteria and Archaea that derive energy by the oxidation of electron donors in the environment. They can be autotrophic or heterotrophic. Chemoautotrophs use inorganic energy sources such as ferrous iron, hydrogen sulphide, elemental sulphur, and ammonia. Chemoheterotrophs can use inorganic energy sources such as sulphur or organic energy sources such as lipids, carbohydrates, and proteins.

Chlorophyll *a*—a specific form of chlorophyll found in all oxygenic photosynthetic organisms. Like other photosynthetic pigments it captures the light energy needed to drive photosynthesis. Its concentration is used as an indicator of photosynthetic biomass.

Chlorophytes—a group of 'green' algae.

Ciliates—heterotrophic protozoans that are usually covered by ordered rows of hair-like structures called cilia that beat in unison allowing movement. Some mixotrophic species contain symbiotic algae or sequester the plastids of their photosynthetic prey.

Contractile vacuole—an organelle in protozoans that is involved in osmoregulation.

Copepodid—developmental stage of a copepod crustacean. There a number of such stages, each terminated by a shedding of the exoskeleton. This final copepodid stage is the adult.

Cryoconite—derived from Greek, 'cryoconite' means 'cold dust' and refers to fine debris found on the surface of glaciers and ice sheets.

Cryolake—ice-topped ponds on the Dry Valleys glaciers that melt at depth during the summer due to the solar heating of dust at their beds.

Cryoprotectant—a substance such as antifreeze proteins that protects cells from damage at temperatures where water freezes, 0°C in freshwater and below 0°C in saline waters.

Cyanobacteria—a phylum in the Kingdom Bacteria. Cyanobacteria undertake oxygenic photosynthesis and are ubiquitous in extreme environments.

DGGE (denaturing gradient gel electrophoresis)—a molecular fingerprinting method that separates polymerase chain reaction (PCR) generated DNA products. The PCR of environmental DNA can generate templates of differing DNA sequence that represent many of the dominant microbial organisms.

Diatoms—the common name for the group of algae known as the Bacillariophyceae. They are typified by possessing a cell wall impregnated by silica.

Dimictic—lakes that undergo two phases of thermal stratification during the course of a year.

Dinoflagellates—a group of Protozoa classified with the phytoflagellates (Phytomastigophora). However, about half of the species lack photosynthetic pigments and are heterotrophic. Others species are mixotrophic. They are common in sea and lake plankton and are the major constituent of sea ice algae communities.

DOC (dissolved organic carbon)—a collective term for the wide variety of organic molecules that are dissolved in most natural waters. DOC in supraglacial waters and in glacier runoff is unusual compared with other terrestrial environments in that the dissolved organic molecules are relatively small and easily utilized by microbial processes.

Dry Valleys—a row of snow-free valleys in Antarctica located within Victoria Land, west of McMurdo Sound. The region is one of the world's most extreme deserts. The valleys occupy an area of $4800\,km^2$, ~0.03% of the continent, and constitute the largest ice-free region in Antarctica.

Encysted—a protozoan cell that has formed a thick cyst wall around itself allowing it to withstand adverse environmental conditions. The cell undergoes dedifferentiation prior to encystment.

Englacial—within the glacier.

Firnification—the process of glacier ice crystal formation from old snow (or 'firn'). Firnification is caused by grain metamorphosis, aided by melt–refreeze cycles, pressure, and vapour migration in the ageing snow.

FISH (fluorescent *in situ* hybridization)—a cytogenetic technique used to detect and localize the presence or absence of specific DNA sequences on chromosomes. FISH uses fluorescent probes that bind to only those parts of the chromosome with which they show a high degree of sequence similarity.

Flagellates—autotrophic (phytoflagellates) and heterotrophic protozoans that possess one or more whip-like structures for feeding and movement.

Flow fingers—these are equivalent to icicles found within Arctic snow packs. Surface melt does not usually seep into a snow pack in a continuous layer. Instead, water descends into the snow pack along certain flow paths. Should these freeze because the deep snow pack is so cold, flow fingers result.

Foraminiferans—a group of shelled protozoans belonging to the Sub-Phylum Sarcodina that also includes the amoebae. The shells are formed of calcium carbonate and are compartmented and fossil forming.

Fungi—a kingdom of eukaryotic, heterotrophic organisms that includes yeasts, molds, and mushrooms.

Furcilia—the final developmental stage of euphausiid crustaceans (krill).

Heterocysts—cells found in some cyanobacterial species that undertake nitrogen fixation.

Heterotroph—an organism that has fed on a carbon source to maintain, grow, and reproduce itself.

High pressure (HP) drainage system—water at the bed of glaciers can only flow under high pressure in this type of drainage system, whose elements include water films at the bed, linked cavities (pockets of water in topographic depressions interlinked by small, easily throttled flow paths), and water-saturated till.

Hypersaline—of higher salinity than seawater, >35‰.

Hyposaline—of lower salinity than seawater, <35‰.

Ice cap—a dome-shaped or plateau-like glacier that usually exhibits ice flow in all directions from its summit.

Ice sheet—an extensive glacierized region, far greater than an ice cap, and presently exemplified by the Greenland and Antarctic Ice Sheets.

Ice shelf—a sheet of coastal, permanent ice cover that is floating upon the ocean. Ice shelves are usually tens to hundreds of metres thick and may be composed of floating glacier and/or perennial sea ice.

Irradiance—the power of solar radiation per unit area that is incident upon a surface. Solar irradiance is usually expressed as watts per square metre. However, studies of the photosynthetically active radiation (PAR) component of solar irradiance tend to use moles of photons per square metre per second instead.

Kame—a small ridge or mound of sediments deposited at or near the margin of a glacier by meltwater. Ice contact deposits such as deltas can also form, referred to as kame deltas.

Katabatic winds—these winds move downhill, carrying high density air from a higher elevation downslope under the force of gravity. Those flowing from the centre of the Greenland and Antarctic Ice Sheets to the margins become very strong. The air warms during the descent because pressure increases.

Last Glacial Maximum (LGM)—the period when terrestrial ice was at its greatest extent during the last glaciation, between ~26000 and ~19000 years ago.

Lateral moraine—a deposit of boulders, stones, and other debris that forms at the flanks of a glacier as a direct result of glacial debris transfer.

Little Ice Age (LIA)—the cold period between ~1550 AD and ~1850 AD, which included three particularly cold intervals, beginning in approximately 1650, 1770, and 1850, each separated by intervals of slight warming. Temperatures in the northern hemisphere dropped by ~1°C.

Low pressure (LP) drainage system—water in this type of drainage system at the bed of a glacier largely flow under the influence of gravity, although when the flow paths become full periodically, the head of water backing up behind can serve to increase flow. Elements of the low pressure drainage system include channels and tunnels.

Marginal debris—a deposit of boulders, stones, and other debris that forms at the margins of a glacier as a result of glacial debris transfer and slope processes upon adjacent mountain sides.

Mass balance—the difference between accumulation and ablation represents the net mass balance of a glacier. It is usually measured or estimated over a time scale of 1 year or more. Mass balance studies therefore report accumulation, ablation, and net balance as three separate mass balance indices.

Meiofauna—microscopic or small metazoans.

Metazoa—multicelluar animals that include both the invertebrates (animals lacking backbones) and vertebrates (animals with backbones). The size range is enormous, from microscopic (e.g. rotifers) to elephants and whales.

Mixotrophy—meaning mixed feeding, this refers to organisms that are capable of both photosynthesis and heterotrophy.

Nauplius (plural nauplii)—the first developmental stages of copepod crustaceans. There are a successive series of these stages, each terminated by the shedding of the exoskeleton. The final naupliar stage metamorphoses into the first copepodid stage. It is also the first non-feeding developmental stage of euphausiid crustaceans (krill).

Nematodes—a phylum of worms known as the roundworms. The phylum has free-living and parasitic species. They range in size from microscopic to many centimeters in length.

Neoproterozoic—the geological era from 1000 to 542 million years ago. The first multicellular fossil life forms were found in this era, as were several worldwide glaciations.

North Atlantic Oscillation (NAO)—the difference between the atmospheric pressure at sea level between the Icelandic low and the Azores high. It has a strong control on the strength and direction of westerly winds and storm tracks across the North Atlantic.

Nutrient limitation—insufficient nitrogen and phosphorus to support photosynthesis and bacterial production.

Osmoregulation—the regulation of salt concentration within cells. It is usually a term applied to Protozoa.

PCR (polymerase chain reaction)—a technique that amplifies copies of DNA generating many copes of a particular DNA sequence.

Perennial lake ice—ice cover that persists from year to year, typical of the Dry Valleys lakes of Antarctica, but also seen elsewhere in Antarctica (lakes that abut glaciers) and in the High Arctic.

Permafrost—perennially frozen ground. The seasonal thaw layer is referred to as the active layer.

Photosynthetic efficiency—calculated as units of carbon fixed (chlorophyll a)$^{-1}$ h^{-1} µmol m^{-2} s^{-1}.

Photosynthetically active radiation (PAR)—that part of the electromagnetic spectrum with wavelengths between 380 and 760nm (visible light) used as energy by photosynthetic organisms to drive photosynthesis.

Phytoplankton—a generic term for microscopic photosynthetic organisms that float freely in the water column. The phytoplankton include Algae, Cyanobacteria, and photosynthetic protozoans (phytoflagellates).

Protozoa—microscopic single-celled organisms that may be autotrophic, heterotrophic, or mixotrophic. The Protozoa are a sub-kingdom in the Kingdom Protista.

PUFAs—polyunsaturated fatty acids.

Q$_{10}$—a temperature coefficient indicating the rate of change in a reaction with a 10°C increase in temperature. As a general rule an increase of 10°C doubles or trebles the rate of metabolism in organisms, depending on the temperature tolerance range of the organism.

REDOX reactions—reactions in which electrons are swapped, such that a reduced compound becomes oxidized and an oxidizing agent becomes reduced. Microbes are able to utilize the energy released from REDOX reactions to promote their vital processes.

Rotifers—microscopic multicellular organisms (Metazoa) belonging to the Phylum Rotifera.

Common in freshwaters, but also occurring in saline waters.

Rotten ice—the surface glacier ice of the weathering crust.

RUBISCO—an enzyme involved in the Calvin–Benson cycle or the photosynthetic carbon reduction cycle. It catalyses the first major step in the process of carbon fixation.

Scytonemin—a pigment found in Cyanobacteria that provides protection against high solar irradiance levels.

Serpentinization—a low temperature metamorphic process in which basic igneous rocks are oxidized and hydrolysed into serpentine during heating at depth in the presence of water. H$_2$ is released during the reaction.

Snow algae—term applied to phytoflagellates that live on and in snow. Many of them possess pigments that protect against solar radiation damage. These pigments create 'red snow' on snow and glacial surfaces.

Southern Ocean—comprises the southernmost marine waters on the planet, from 60°S latitude to Antarctica. It is the fourth largest of the five principal oceanic divisions, and is a region that is critical in controlling worldwide atmospheric CO$_2$ concentrations.

Stable isotope—isotopes are elements that have the same number of protons but different numbers of neutrons in their nucelii. Stable isotopes are those that do not partake in radiochemical reactions, and therefore remain in their present state. ^{13}C and ^{12}C are stable isotopes of carbon, whereas ^{14}C is radiochemical and decays over time.

Subglacial—at the ice–bed interface beneath a glacier.

Subglacial lake—a lake at the bed of a glacier.

Sublimation—a phase transition of snow and ice directly to vapour. Sublimation is a major ablation process from the surface of cold polar ice.

Superimposed ice—a refrozen meltwater layer that forms on the surface of a cold ice layer, usually beneath a melting snow pack. Refrozen snow melt is a major contributor to surface ice accumulation in the Arctic, and it occurs at the beginning and end of the summer melt period.

Supraglacial—upon the glacier surface.

Supraglacial lake—a lake on the surface of a glacier.

Temperature gradient metamorphism (TGM)—the growth of new snow crystals within snow packs as brought about by temperature gradients of

~20°C m^{-1}. Snow crystals formed under this regime are poorly bonded together, so the snow cover is prone to avalanche.

Turbellarians—so-called flat worms belonging to the Phylum Playthelminthes.

Ultramicrobacteria—strains of bacteria that are inherently small (0.2–0.3μm long). Small size also results in bacteria that are starved.

Ultraviolet light—electromagnetic radiation in the spectral band between 10 and 400nm. The upper range is usually described as either UV-A (400–320nm) or UV-B (320–290nm). These represent about 6% and 2% of clear sky solar irradiance, respectively, the rest being dominated by visible light (42%) and infrared (about 50%). UV light is important due to its strong absorbance by biological materials, causing damage to cells and fluorescence emission.

Valley glacier—a topographically constrained glacier within a valley and therefore of different morphology to either an ice cap or ice sheet.

Viruses—possess either DNA or RNA. Their shape varies from simple helical and icosahedral forms to more complex structures. They cannot reproduce and can only replicate by parasitizing a plant, animal, or bacterial cell which then becomes a virus-producing factory.

Weathering crust—the distinctive white, low density ice layer that forms on the surface of glaciers when ablation is predominantly due to solar irradiance. Weathering crusts that are tens of centimetres thick can form during sunny conditions, only to be replaced by hard, solid ice surfaces during rain and periods when sensible heat transfer over the glacier surface exceeds that provided by solar radiation.

References

Abyzov, S.S. (1993). Microorganisms in Antarctic ice. In Friedmann, E. (ed.) *Antarctic Microbiology*, pp 265–95. Wiley Liss, New York.

ACERE (2009). *Transition and Tipping Points in Complex Environmental Systems. A report by the National Science Foundation Advisory Committee for Environmental Research and Education.* ACERE, Washington, DC, 56 pp.

ACIA (2005). *Arctic Climate Impact Assessment—Scientific Report.* Cambridge University Press, Cambridge, UK, 1046 pp.

Alfreider, A., Pernhaler, J., Amann, R., et al. (1996). Community analysis of the bacterial assemblages in the winter cover and pelagic layers of a high mountain lake by in situ hybridization. *Applied and Environmental Microbiology* **62**, 2138–44.

Alger, A.S., McKnight, D.M., Spaulding, S.A., *et al.* (1997). *Ecological Processes in a Cold Desert Ecosystem: The abundance and species distribution of algal mats in glacial meltwater streams in Taylor valley, Antarctica.* Institute of Arctic and Alpine Research Occasional Paper No. 51. University of Colorado, Boulder, CO.

Alldredge, A.L. and Silver, M.W. (1988). Characteristics, dynamics and significance of marine snow. *Progress in Oceanography* **20**, 41–82.

Amato, P., Hennebelle, R., Magand, O., *et al.* (2007). Bacterial characterization of the snow cover at Spitzberg, Svalbard. *FEMS Microbiology* **59**(2), 255–64.

Amon, R.M.W. and Benner, R. (1996). Bacterial utilization of different size classes of dissolved organic carbon. *Limnology and Oceanography* **41**, 41–51.

Anderson, J.D., Schubert, G., Jacobson, R.A., Lau, E.L., Moore, W.B. and Sjogren, W.L. (1998). Europa's differentiated internal structure: inferences from four Galileo encounters. *Science* **281**, 2019–22.

Anderson, S.P., Drever, J.I. and Humphrey, N.F. (1997). Chemical weathering in glacial environments. *Geology* **25**, 399–402.

Anesio, A.M., Mindl, B., Laybourn-Parry, J. and Sattler, B. (2007). Virus dynamics on a high Arctic glacier (Svalbard). *Journal of Geophysical Research* **112**(G04S31), doi 10.1029/2006JG000350.

Anesio, A.M., Hodson, A.J., Fritz, A., Psenner, R. and Sattler, B. (2009). High microbial activity on glaciers: importance to the global carbon cycle. *Global Change Biology* **15**, 955–60.

Anesio, A.M., Sattler, B., Foreman, C., Hodson, A.J., Tranter, M. and Psenner, R. (2011). Carbon fluxes through bacterial communities on glacier surfaces. *Annals of Glaciology* **51**(56), 32–40.

Archer, S.D., Leakey, R.J.G., Burkill, P.H., Sleigh, M.A. and Appleby, C.J. (1996). Microbial ecology of sea ice at a coastal Antarctic site; community composition, biomass and temporal change. *Marine Ecology Progress Series* **135**, 179–95.

Arndt, C.E. and Swadling, K.M. (2006). Crustacea in Arctic and Antarctic sea ice; distribution, diet and life histories. *Advances in Marine Biology* **51**, 197–315.

Arrigo, K.R. and van Dijken, G.L. (2003). Phytoplankton dynamics within 37 Antarctic coastal polynya systems. *Journal of Geophysical Research* **108**(3271), doi 10.1029/2002JC002739.

Arrigo, K., Worthen, D.L., Dixon, P. and Lizotte, M.P. (1998). Primary productivity of near surface communities within Antarctic pack ice. In Lizotte, M.P. and Arrigo, K.R. (eds) *Antarctic Sea Ice Biological Processes, Interactions and Variability*, pp 23–43. Antarctic Research Series No. 73. American Geophysical Union, Washington, DC.

Arrigo, K.R., Pabi, S., van Dijken, G.L. and Maslowski, W. (2010). Air–sea flux of CO_2 in the Arctic Ocean, 1998–2003. *Journal of Geophysical Research* **115** (G04024), doi (10.1029/2009jg001224).

Atreya, S.K., Witasse, O., Chevier, V.F., *et al.* (2011). Methane on Mars: current observations, interpretation, and future plans. *Planetary and Space Science* **59**, 133–6.

Azam, F., Fenchel, T., Field, J.G., Gray, J.S., Meyer-Reil, R.A. and Thingstad, F. (1983). The ecological role of water column microbes in the sea. *Marine Ecology Progress Series* **6**, 213–20.

Bagshaw, E.A., Wadham, J.L., Mowlem, M., *et al.* (2010). Determination of dissolved oxygen in the cryosphere: a comprehensive laboratory and field evaluation of fibre optic sensors. *Environmental Science and Technology* **45**, 700–5.

Bagshaw, L., Tranter, M., Wadham, J.L., Fountain, A.G. and Mowlem, M. (2011). High-resolution monitoring reveals dissolved oxygen dynamics in an Antarctic cryoconite hole. *Hydrological Processes* **18**, 2868–77.

Baker, V.R. (1982). *The Channels of Mars*. 198 pp. University of Texas, Austin, TX.

Baker, V.R., Dohm, J.M., Fairen, A.G., *et al.* (2005). Extraterrestrial hydrogeology. *Hydrogeology Journal* **13**, 51–68.

Bales, R.C., Davis, R.E. and Stanley, D.A. (1989). Ion evolution through shallow homogenous snowpack. *Water Resources Research* **25**, 1869–77.

Bales, R.C., Davis, R.E. and Williams, M.W. (1993). Tracer release in melting snow—diurnal and seasonal patterns. *Hydrological Processes* **7**, 389–401.

Banks, M.E., Lang, N.P., Kargel, J.S., *et al.* (2008). Analysis of sinous ridges in the southern Argyre Planitia, Mars using HiRISE and CTX images and MOLA data. *Journal of Geophysical Research* **114**(E09003), doi 10.1029/2008JE 003244.

Barker, J.D., Sharp, M.J. and Turner, R.J. (2009). Using synchronous fluorescence spectrometry and principal components analysis to monitor dissolved organic matter dynamics in a glacier system. *Hydrological Processes* **23**, 1487–500.

Barker, J.D., Klassen, J.L., Sharp, M.J., Fitzsimmons, S.J. and Turner, R.J. (2010). Detecting biogeochemical activity in basal ice using fluorescence spectroscopy. *Annals of Glaciology* **56**, 47–55.

Barrett, P. (1999). *How Old is Lake Vostok?* Scientific Committee for Antarctic Research International Workshop on Subglacial Lake Exploration No. 2. Scot Polar Research Institute, Cambridge, UK, 17 pp.

Bartholomew, I., Nienow, P., Mair, D., Hubbard, A., King, M.A. and Sole, A. (2010). Seasonal evolution of subglacial drainage and acceleration in a Greenland outlet glacier. *Nature Geosciences* **3**, 408–11.

Battin, T.J., Wille, A., Sattler, B. and Psenner, R. (2001). Phylogenetic and functional heterogeneity of sediment biofilms along environmental gradients in a glacial stream. *Applied and Environmental Microbiology* **67**, 799–807.

Bauer, H., Kasper-Giebl, A., Löflund, M., *et al.* (2002). The contribution of bacteria and fungal spores to organic carbon content of cloud water, precipitation and aerosols. *Atmospheric Research* **64**, 109–19.

Beatty, J.T., Overmann, J., Lince, M.T., et al (2005). An obligately photosynthetic bacterial anaerobe from a deep-sea hydrothermal vent. *Proceedings of the National Academy of Sciences* **102**, 9306–10.

Bell, R.E. (2008). The role of subglacial water in ice-sheet mass balance. *Nature Geoscience* **1**, 297–304.

Benn, D.I. and Evans, D.J.A. (2010). *Glaciers and Glaciation*, 2nd edn. Arnold, London, 734 pp.

Benner, S.A. (2010). Defining life. *Astrobiology* **10**, 1021–30.

Berggren, M., Laudon, H., Haei, M., Strom, L. and Jansson, M. (2009). Efficient aquatic bacterial metabolism of dissolved low-molecular-weight compounds from terrestrial sources. *ISME Journal* **4**, 408–16.

Bhartia, R., Bhatia, M.P., Hug, W.F., *et al.* (2008). Classification of organic and biological materials with deep ultraviolet excitation. *Applied Spectroscopy* **62**, 1070–7.

Bhatia, M., Sharp, M. and Foght, J. (2006). Distinct bacterial communities exist beneath a high Arctic polythermal glacier. *Applied and Environmental Microbiology* **72**, 5838–45.

Bhatia, M., Das, S.B., Longnecker, K., Charette, M.A. and Kujawinski, E.B. (2010). Molecular characterization of dissolved organic matter associated with the Greenland Ice Sheet. *Geochimica et Cosmochimica Acta* **74**, 3768–84.

Bintanja, R. (1999). On the glaciological, meteorological and climatological significance of Antarctic blue ice areas. *Reviews of Geophysics* **37**(3), 337–59.

Black, R.F., Jackson, M.L. and Berg, T.E. (1965). Saline discharge from Taylor Glacier, Victoria Land, Antarctica. *Journal of Geology* **74**, 175–81.

Blackford, J.R. (2007). Sintering and microstructure of ice: a review. *Journal of Physics D—Applied Physics* **40**, R355–85.

Bluhm, B.A. and Gradinger, R. (2008). Regional variability in food availability for Arctic marine mammals. *Ecological Applications* **18**, S77–S96.

Bøggild, C.E., Brandt, R.E., Brown, K.J. and Warren, S.G. (2010). The ablation zone in northeast Greenland: ice types, albedos and impurities. *Journal of Glaciology* **56**(195), 101–13.

Bondarenko, N.A., Timoshkin, O.A., Röpstorf, P. and Melnik, N.G. (2006). The under-ice and bottom periods in the life cycle of *Aulacoseira baicalensis* (K. Meyer) Simonsen, a principal Lake Baikal alga. *Hydrobiologia* **568**, 107–9.

Borriss, M., Helmke, E., Hanschke, R. and Schweder, T. (2003). Isolation and characterization of marine psychrophylic phage-host systems from Arctic sea ice. *Extremophiles* **7**, 377–84.

Bottrell, S.H. and Tranter, M. (2002). Sulphide oxidation under partially anoxic conditions at the bed of the Haut Glacier d'Arolla, Switzerland. *Hydrological Processes* **16**, 2363–8.

Bowling, D.R., Massmna, W.J., Schaeffer, S.M., Burns, S.P., Monson, R.K. and Williams, M.W. (2009). Biological and physical influences on the carbon isotope content of CO_2 in a subalpine forest snowpack, Niwot Ridge, Colorado. *Biogeochemistry* **95**, 37–59.

Box, J.E. and Ski, K (2007). Remote sounding of Greenland supraglacial melt lakes: implications for subglacial hydraulics. *Journal of Glaciology* **53**, 181.

Boynton, W.V., Feldman, W.C., Squyres, S.W., *et al.* (2002). Distribution of hydrogen in the near surface of Mars: evidence for subsurface ice deposits. *Science* **297**, 81–5.

Brinkmeyer, R., Knittel, K., Jürgens, J., Weyland, H., Amann, R. and Helmke, E. (2003). Diversity and structure of bacterial communities in Arctic versus Antarctic pack ice. *Applied and Environmental Microbiology* **69**, 6610–19.

Broecker, W.S. (1985). *How to Build a Habitable Planet*. Eldigio Press, New York, 291 pp.

Brooks, P.D., McKnight, D.M. and Elder, K. (2004). Carbon limitation of soil respiration under winter snowpacks: potential feedbacks between growing season and winter carbon fluxes. *Global Change Biology* **11**, 231–8.

Bulat, S.A., Alekhina, I.A., Blot, M., *et al.* (2004). DNA signature of thermophilic bacteria from the aged accretion ice of Lake Vostok, Antarctica: implications for searching for life in extreme icy environments. *International Journal of Astrobiology* **3**, 1–12.

Burr, D.M., McEwen, A.S. and Sakimoto, S.E.H. (2002a). Recent aqueous floods from the Cerberus Fossae, Mars. *Geophysical Research Letters* **29**, doi 10.1029/2001GL013345.

Burr, D.M., Grier, A.S., McEwen, A.S. and Keszthelyi, L.P. (2002b). Repeated aqueous flooding from the Cerberus Fossae: evidence for very recent extant, deep groundwater on Mars. *Icarus* **159**, 53–73.

Burrow, S.G., Wadham, J.L., Salter, M.A. and Barnes, R. (2009). E-tracers—a new technique for glacial sensing. American Geophysical Union Fall Meeting, San Francisco, Abstract C43B–0506.

Byre, S. and Ingersoll, A.P. (2003). A sublimation model for Martian South Pole ice features. *Science* **299**, 1051–3.

Cameron, K.A. (2010). *Microbial community structure, diversity and biogeochemical cycling in Arctic and Antarctic cryoconite holes*. Unpublished PhD thesis, University of Sheffield, Sheffield, UK, 241 pp.

Carey, A.G. and Montagna, P.A. (1982). Arctic sea ice faunal assemblage; first approach to description and source of the underice meiofauna. *Marine Ecology Progress Series* **8**, 1–8.

Carlson, R.W., Anderson, M.S., Johnson, R.E., *et al.* (1999). Hydrogen peroxide on the surface of Europa. *Science* **283**, 2062–4.

Carlson, R.W., Anderson, M.S., Johnson, R.E., Schulman, M.B. and Yavrouian, A.H. (2002). Sulfuric acid production on Europa: the radiolysis of sulphur in water ice. *Icarus* **157**, 456–63.

Caron, D.A. (1987). Grazing of attached bacteria by heterotrophic microflagellates. *Microbial Ecology* **13**, 203–18.

Carpenter, E.J. and Capone, D.G. (2003). Authors' reply to a letter to the editor. *Applied and Environmental Microbiology* **69**, 6340–1.

Carpenter, E.J., Lin, S. and Capone, D.G. (2000). Bacterial activity in South Pole snow. *Applied and Environmental Microbiology* **66**, 4514–17.

Carr, M.H. and Head, J.W. (2003). Oceans on Mars: an assessment of the observational evidence and possible fate. *Journal of Geophysical Research* **108**, doi 10.1029/2002JE001963.

Castello, J. (ed.) *Life in Ancient Ice*, pp 209–27. Princeton University Press, Princeton, NJ.

Castillo-Rogez, J.C., Matson, D.L., Vance, S.D., Davies, A.G. and Johnson, T.V. (2007). *The early history of Enceladus: setting the scene for today's activity*. Lunar and Planetary Science Meeting, Houston, Abstract No. 2265.

Catalan, J. (1989). The winter cover of a high-mountain Mediterranean lake (Estany Redó, Pyrenees). *Water Resources Research* **25**, 519–27.

Charlebois, R.L., Beiko, R.G. and Ragan, M.A. (2003). Microbial phylogenetics: branching out. *Nature* **421**, 271.

Chen, S. and Baker, I. (2010). Evolution of individual snowflakes during metamorphism. *Journal of Geophysical Research* **115**(D21114), doi 10.1029/2010JD014132.

Chernoff, D.I. and Bertram, A.K. (2010). Effects of sulfate coatings on the ice nucleation properties of a biological ice nucleus and several types of minerals. *Journal of Geophysical Research* **115**(D21114), doi 10.1029/2010JD014132.

Christner, B.C., Mosley-Thompson, E., Thompson, L. and Reeve, J.N. (2001). Isolation of bacteria and 16S rDNAs from Lake Vostok accretion ice. *Environmental Microbiology* **3**, 570–7.

Christner, B.C., Kvitko, B.H. and Reeve, J.N. (2003). Molecular identification of bacteria and Eukarya inhabiting an Antarctic cryoconite hole. *Extremophiles* **7**, 177–83.

Christner, B.C., Mosley-Thompson, E., Thompson, L.G. and Reeve, J.N. (2005a). Recovery and identification of bacteria from polar and non-polar glacial ice. In Rogers, S.O. and Castello, J. (eds) *Life in Ancient Ice*, pp 209–27. Princeton University Press, Princeton, NJ.

Christner, B.C., Mikucki, J.A., Foreman, C.M., Denson, J. and Priscu, J.C. (2005b). Glacial ice cores: a model system for developing extraterrestrial decontamination protocols. *Icarus* **174**, 572–84.

Christner, B.C., Royston-Bishop, G., Foreman, C.M., *et al.* (2006). Limnological conditions in subglacial Lake Vostok. *Limnology and Oceanography* **51**, 2485–501.

Christner, B.C., Morris, C.E., Foreman, C.M., Cai, R. and Sands, D.C. (2008). Ubiquity of biological ice nucleators in snowfall. *Science* **319**, 1214.

Chróst, T.J., Münster, U., Rai, H., Albrecht, D., Witzel, K.P. and Overbeck, J. (1989). Photosynthetic production and exoenzymatic degradation of organic matter in the euphotic zone of a eutrophic lake. *Journal of Plankton Research* **11**, 223–42.

Chyba, C.F. (2000). Energy for microbial life on Europa. *Nature* **403**, 381–2.

Chyba, C.F. and Hand, K.P. (2001). Planetary science—life without photosynthesis. *Science* **292**, 2026–7.

Clark, A. and Leakey, R.J.G. (1996). The seasonal cycle of phytoplankton, macronutrients and the microbial community in a nearshore Antarctic marine ecosystem. *Limnology and Oceanography* **41**, 1281–94.

Clark, B., Baird, A., Weldon, R., Tsusaki, D., Schnabel, L. and Candelaria, M. (1982). Chemical composition of Martian fines. *Journal of Geophysical Research* **87**, 10059–68.

Clark, P.U., Dyke, A.S., Shakun, J.D., *et al.* (2009). The Last Glacial Maximum. *Science* **325**, 710–14.

Clow, G.D., Saltus, R.W. and Waddington, E.D. (1996). A new, high precision borehole-temperature logging system used at GISP2, Greenland and Taylor Dome, Antarctica. *Journal of Glaciology* **42**, 576–84.

Cockell, C.S., Bagshaw, E., Balme, M., *et al.* (2011). Subglacial environments and the search for life beyond Earth. In Siegert, M.J., Kennicutt, M.C. and Bindchadler, R.A.) *Antarctic Subglacial Aquatic Environments*, pp 129–48. Geophysical Monograph Series, Vol. 192. American Geophysical Union, Washington, DC.

Coffin, R.B. (1989). Bacterial uptake of dissolved free and combined amino acids in estuarine waters. *Limnology and Oceanography* **34**, 531–42.

Coker, J.A., Sheridan, P.P., Loveland-Curtze, J., Gutshall, K.R., Auman, A.J. and Brenchley, J.E. (2003). Biochemical characterization of a β-galactosidase with a low temperature optimum obtained from an Antarctic *Arthrobacter* isolate. *Journal of Bacteriology* **185**, 5473–82.

Colbeck, S.C. (1991). The layered character of snow covers. *Reviews of Geophysics* **29**, 81–96.

Cook, J., Hodson, A.J., Telling, J., Anesio, A., Irvine-Fynn, T. and Bellas, C. (2010). The mass–area relationship within cryoconite holes and its implications for primary production. *Annals of Glaciology* **51**, 106–10.

Cooper, P.D., Johnson, R.E., Mauk, B.H., Garrett, H.B. and Gehrels, N. (2001). Energetic ion and electron irradiation of the icy Galilean statellites. *Icarus* **149**, 133–59.

Cooper, P.D., Johnson, R.E. and Quickenden, T.I. (2003). Hydrogen peroxide dimmers and production of O_2 in icy satellite surfaces. *Icarus* **166**, 444–6.

Corbisier, T.N., Petti, M.A.V., Skowronski, R.S.P. and Brito, T.A.S. (2004). Trophic relationships in the nearshore zone of Martel Inlet (King George Island, Antarctica): δ ^{13}C stable-isotope analysis. *Polar Biology* **27**, 75–82.

Cragin, J.H., Hewitt, A.D. and Colbeck, S.C. (1996). Grain-scale mechanisms influencing the elution of ions from snow. *Atmospheric Environment* **30**, 119–27.

Dalton, J.B., Mogel, R., Kagawa, H.K., Chan, S.L. and Jamieson, C.S. (2003). Near-infrared detection of potential evidence for microscopic organisms on Europa. *Astrobiology* **3**, 505–29.

Das, S.B., Joughin, I., Behn, M.D., *et al.* (2008). Fracture propagation to the base of the Greenland Ice Sheet during supraglacial lake drainage. *Science* **320**(5877), 778–81.

Davies, T.D., Abrahams, P.W., Tranter, M., Blackwood, I., Brimblecombe, P. and Vincent, C.E. (1984). Black acidic snow in remote Scottish Highlands. *Nature* **312**, 58–61.

Davies, T.D.,Tranter, M. and Jones, H.G. (eds) (1991). *Seasonal Snowpacks: Processes of compositional change.* NATO ASI Series G, Ecological Sciences Vol. 28. Springer-Verlag, Berlin, 471 pp.

Davies, T.D., Tranter, M., Jickells, T.D., *et al.* (1992). Heavily-contaminated snowfalls in the remote Scottish Highlands—a consequence of regional-scale mixing and transport. *Atmospheric Environment* **26**, 95–112.

De Angelis, M., Petit, J.R., Savarino, J., Souchez, R. and Thiemens, M.H. (2004). Contributions of an ancient evaporitic-type reservoir to subglacial lake Vostok chemistry. *Earth and Planetary Science Letters* **22**, 751–65.

De la Mare, W.K. (2009). Changes in Antarctic sea-ice extent from direct historical observations and whaling records. *Climatic Change* **92**, 461–93.

De Mora, S.J., Whitehead, R.F. and Gregory, M. (1994). The chemical composition of glacial meltwater ponds and streams on the McMurdo Ice Shelf, Antarctica. *Antarctic Science* **6**, 17–27.

Debenham, F. (1920). A new mode of transportation by ice: the raised marine muds of South Victoria Land (Antarctica). *Quarterly Journal of the Geological Society of London* **75**, 51–76.

del Giorgio, P.A., Cole, J.J. and Cimbleris, A. (1997). Respiration rates in bacteria exceed phytoplankton production in unproductive aquatic systems. *Nature* **385**, 148–51.

Devos, N., Ingouff, M., Loppes, R. and Matagne, R.F. (1998). RUBISCO adaptation to low temperatures: a comparative study in psychrophilic and mesophilic unicellular algae. *Journal of Phycology* **34**, 655–60.

Dhaked, R.K., Alam, S.I., Dixit, A. and Singh, L. (2005). Purification and characterization of thermo-labile alkaline phosphatase from an Antarctic psychrotolerant *Bacillus* sp. P9. *Enzyme Microbiology and Biotechnology* **36**, 855–61.

Dobson, S.J., Colwell, R.R., McMeekin, T.A. and Franzmann, P.D. (1993). Direct sequencing of the polymerase chain reaction-amplified 16S ribosomal RNA gene of *Flavobacterium salegens* sp. nov. 2 new species from a hypersaline Antarctic lake. *International Journal of Systematics and Bacteriology* **43**, 77–83.

Doran, P.T., McKay, C.P., Adams, W.P., English, M.C., Wharton, R.A. and Meyer, M.A. (1996). Climate forcing and thermal feedback of residual ice-covers in the high Arctic. *Limnology and Oceanography* **41**, 839–48.

Doran, P.T., Fritsen, C.H., McKay, C.P., Priscu, J.C. and Adams, E.E. (2003). Formation and character of an ancient 19-m ice cover and underlying trapped brine in an 'ice-sealed' east Antarctic lake. *Proceedings of the National Academy of Sciences* **100**, 26–31.

Doran, P.T., Fritsen, C.H., Murray, C.H., Kenig, F., McKay, C.P. and Kyne, J.D. (2008). Entry approach into pristine ice-sealed lakes—Lake Vida, East Antarctica, a model ecosystem. *Limnology and Oceanography Methods* **6**, 542–7.

Doran, P.T., Lyons, W.B. and McKnight, D.M. (eds) (2010). *Life in Antarctic Deserts and Other Cold Dry Environments*. Cambridge University Press, Cambridge, UK.

Dowdeswell, J.A. and Siegert, M.J. (1999). The dimensions and topographic setting of Antarctic subglacial lakes and implications or large scale water storage beneath continental ice sheets. *Geological Society of America Bulletin* **111**, 254–63.

Dubnick, A., Barker, J., Wadham, J.L., *et al.* (2010). Characterization of dissolved organic matter (DOM) from glacial environments using total fluorescence spectroscopy and parallel factor analysis. *Annals of Glaciology* **56**, 111–22.

Ducklow, H.W. (1983). Production and fate of bacteria in the ocean. *Bioscience* **33**, 494–501.

Duguay, C.R., Prowse, T.D., Bonsal, B.R., Brown, R.D., Lacroix, M.P. and Ménard, P. (2006). Recent trends in Canadian ice cover. *Hydrological Processes* **20**, 781–801.

Dumont, I., Schoemann, V., Lannuzel, D., Chou, L., Tilson, J-L. and Becquevort, S. (2009). Distribution and characterization of dissolved and particulate organic carbon in Antarctic pack ice. *Polar Biology* **32**, 733–50.

Duval, B., Shetty, K. and Thomas, W.H. (2000). Phenolic compounds and antioxidant properties in the snow alga *Chlamydomonas nivalis* after exposure to UV light. *Journal of Applied Phycology* **11**, 559–66.

Dyurgerov, M. (2002). *Glacier Mass Balance and Regime: Data of measurements and analysis*. Institute of Arctic and Alpine Research Occasional Paper No. 55. University of Colorado, Boulder, CO, 91 pp.

Dyurgerov, M.B. and Meier M.F. (2005). *Glaciers and the Changing Earth System: A 2004 snapshot*. Institute of Arctic and Alpine Research Occasional Paper No. 58. University of Colorado, Boulder, CO.

Dyurgerov, M., Bring, A. and Destouni, G. (2010). Integrated assessment of changes in freshwater inflow to the Arctic Ocean. *Journal of Geophysical Research* **115** (D12116), doi 10.1029/2009jd013060.

Edwards A., Anesio A.M., Rassner S.M., *et al.* (2010). Possible interactions between bacterial diversity, microbial activity and supraglacial hydrology of cryoconite holes in Svalbard. *ISME Journal* **5**, 150–60.

Ellis-Evans, J.C. and Wynn-Williams, D. (1996). Antarctica: a great lake under the ice. *Nature* **381**, 644–6.

Escobar, F., Vidal, F., Garin, C. and Naruse, R. (1992). Water balance in the Patagonian icefield. In Naruse, R. and Aniya, M. (eds) *Glaciological Researches in Patagonia 1990*, pp 109–11. Japanese Society of Snow and Ice, Nagoya, Japan.

Fahnestock, M., Abdalati, W., Joughin, I., Brozena, J. and Gogineni, P. (2001). High geothermal heat flow, basal melt and the origin of rapid ice low in central Greenland. *Science* **294**, 2338–42.

Falkowski, P.G. and Raven, J.A. (1997). *Aquatic Photosynthesis*. Blackwell Science, Malden, MA.

Fanale, F.P. (1976). Martian volatiles: their degassing history and geochemical fate. *Icarus* **28**, 179–202.

Fanale, F.P., Postawko, S.E., Pollack, J.B., Carr, M.H. and Pepin, R.O. (1992). Mars: epochal climate change and volatile history. In Kieffer, H.H. (ed.) *Mars*, pp 1135–79. University of Arizona Press, Tucson, AZ.

Fantoni, R., Barbini, R., Colao, F., Ferrante, D., Fiorani, L. and Palucci, A. (2004). Integration of two lidar fluorosensor payloads in submarine ROV and flying UAV platforms. *EARel eProceedings* **3**, 1/2004 43–53.

Feldman, W.C., Prettyman, T.H., Maurice, S., *et al.* (2004). Global distribution of near-surface hydrogen on Mars. *Journal of Geophysical Research* **109** (E09000), doi 10.1029/2003JE002160.

Felip, M., Sattler, B., Psenner, R. and Catalan, J. (1995). Highly active microbial communities in the ice and snow cover of high mountain lakes. *Applied and Environmental Microbiology* **61**, 2394–401.

Felip, M., Camarero, L. and Catalan, J. (1999). Temporal changes of microbial assemblages in the ice and snow cover of a high mountain lake. *Limnology and Oceanography* **44**, 973–87.

Fellman, J.B., Spencer, R.G.M., Hernes, P.J., Edwards, R.T., D'Amore, D.V. and Hood, E. (2010). The impact of glacier runoff on the biodegradability and biochemical composition of terrigenous dissolved organic matter in near-shore marine ecosystems. *Marine Chemistry* **121**(1–4), 112–22.

Fernández-Valiente, E., Quesada, A., Howard-Williams, C. and Hawes, I. (2001). N_2-fixation in cyanobacterial mats from ponds on the McMurdo Ice Shelf, Antarctica. *Microbial Ecology* **42**, 338–49.

Fettweis, X., van Ypersele, J-P., Gallée, H., Lefebre, F. and Lefebvre, W. (2007). The 1979–2005 Greenland ice sheet melt extent from passive microwave data using an improved version of the melt retrieval XPGR algorithm. *Geophysical Research Letters* **34**, L05502, doi 10.1029/2006GL028787.

Fischer, H., Wagenbch, P. and Kipfstuhl, J. (1998). Sulfate and nitrate firn concentrations on the Greenland ice sheet—2 temporal anthropogenic deposition changes. *Journal of Geophysical Research—Atmospheres* **103**, 21935–42.

Fischer, U.R. and Velimirov, B. (2002). High control of bacterial production by viruses in a eutrophic oxbow lake. *Aquatic Microbial Ecology* **27**, 1–12.

Fleischmann, R.D., Adams, M.D., White, O., *et al.* (1995). Whole-genome random sequencing and assembly of *Haemophilus influenza*. *Science* **269**, 496–512.

Fogg, G.E. (1967). Observations on the snow algae of the South Orkney Islands. *Proceedings of the Royal Society London, Series B* **252**, 279–87.

Fogg, G.E. (1988). *The Biology of Polar Habitats.* Oxford University Press, Oxford, UK.

Foght, J., Aislabie, J., Turner, S., *et al.* (2004). Culturable bacteria in subglacial sediments and ice from two Southern Hemisphere glaciers. *Microbial Ecology* **47**, 329–40.

Foissner, W., Berger, H. and Schaumburg, J. (1999). *Identification and Ecology of Limnetic Plankton Ciliates.* Bayerisches Landesamt für Wasserwirtschafft, Munich.

Folk, R.L. (1993). SEM imaging of bacteria and nannobacteria in carbonate sediments and rocks. *Journal of Sedimentary Research* **63**, 990–9.

Foreman, C.M., Wolf, C.F. and Priscu, J.C. (2004), Impact of episodic warming events on the physical, chemical and biological relationships of lakes in the McMurdo Dry Valleys, Antarctica. *Aquatic Geochemistry* **10**, 239–68.

Foreman, C.M., Sattler, B., Mikucki, J.A., Porazinska, D.L. and Priscu, J.C. (2007). Metabolic activity and diversity of cryoconites in the Taylor Valley, Antarctica. *Journal of Geophysical Research* **112**(G04S32), doi 10.1029/2006JG000358.

Formisano, V., Atreya, S., Encrenaz, T., Ignatiev, N. and Giuranna, M. (2004). Detection of methane in the atmosphere of Mars. *Science* **306**, 1758–61.

Fornea, A.P., Brooks, S.D., Dooley, J.B. and Saha, A. (2009). Heterogeneous freezing of ice on atmospheric aerosols containing ash, soot and soil. *Journal of Geophysical Research* **114**(D13201), doi 10.1029/2009JD011958.

Fortes, A.D. (2000). Exobiological implications of a possible ammonia–water ocean inside Titan. *Icarus* **146**, 144–52

Fortner, S.K., Tranter, M., Fountain, A.G., Lyons, B. and Welch, K.A. (2005). The geochemistry of supraglacial streams of Canada Glacier, Taylor Valley (Antarctica), and their evolution into proglacial waters. *Aquatic Geochemistry* **11**, 391–412.

Foti, M., Piattelli, M., Amico, V. and Ruberto, G. (1994). Antioxidant activity of penolic meroditerpenoids from marine algae. *Journal of Photochemistry and Photobiology B Biology* **26**, 159–64.

Fountain, A.G. (1996). Effect of snow and firn hydrology on the physical and chemical characteristics of glacial runoff. *Hydrological Processes* **10**, 509–21.

Fountain, A.G. and Tranter, M. (2008). Introduction to special section on microcosms in ice: the biogeography of cryoconite holes. *Journal of Geophysical Research* **113**(G02S91), doi 10.1029/2008JG000698.

Fountain, A.G. and Walder, J.S. (1998). Water flow through temperate glaciers. *Reviews of Geophysics* **36**, 299–328.

Fountain, A.G., Tranter, M., Nylen, T.H., Lewis, K.L. and Mueller, D.R. (2004). Evolution of cryoconite holes and their contribution to meltwater runoff from glaciers in the McMurdo Dry Valleys, Antarctica. *Journal of Glaciology* **50**(168), 35–45.

Frenette, J-J., Thibeault, P. and Lapierre, J-F. (2008). Presence of algae in freshwater ice cover of fluvial Lac Saint-Pierre (St Lawrence River, Canada). *Journal of Phycology* **44**, 284–91.

Fricker, H.A., Scambos, T., Carter, S., *et al.* (2010). Synthesising multiple remote sensing techniques for subglacial hydrologic mapping: application to a lake system beneath MacAyeal Ice Stream, Antarctica. *Journal of Glaciology* **56**, 187–99.

Fritsen, C.H. and Priscu, J.C. (1998). Cyanobacterial assemblages in permanent ice covers on Antarctic lakes: distribution, growth rate, and temperature response of photosynthesis. *Journal of Phycology* **34**, 587–97.

Fritsen, C.H., Lytle, V.I., Ackley, S.F. and Sullivan, C.W. (1994). Autumn bloom of Antarctic pack-ice algae. *Science* **266**, 782–4.

Fritsen, C.H., Ackley, S.F., Kremer, J.N. and Sullivan, C.W. (1998). Flood–freeze cycles and microalgal dynamics in Antarctic pack ice. In Lizotte, M.P. and Arrigo, K.R. (eds) *Antarctic Sea Ice Biological Processes, Interactions and Variability*, pp 1–21. Antarctic Research Series No. 73. American Geophysical Union, Washington, DC.

Fukushima, H. (1963). Studies on cryptophytes in Japan. *Journal of the Yokohama Municipal University C* **43**, 1–146.

Fuhrman, J. (1987). Close coupling between release and uptake of dissolved free amino acids in seawater studied isotope dilution approach. *Marine Ecology Progress Series* **37**, 45–52.

Gaidos, E., Nealson, K.H. and Kirschvink, J.L. (1999). Life in ice-covered oceans. *Science* **284**, 1631.

Gaidos, E., Lanoil, B., Thorsteinsson, T., *et al.* (2004). A viable microbial community in a subglacial volcanic crater lake, Iceland. *Astrobiology* **4**, 327–34.

Gaidos, E., Marteinson, V., Thorsteinsson, T., *et al.* (2008). An oligarchic microbial assemblage in the anoxic bottom waters of a volcanic subglacial lake. *ISME Journal* **4**, 486–97.

Garrison, D.L. and Buck, K.R. (1991). Sea ice assemblages in Antarctic pack ice during the austral spring: environmental conditions, primary production and community structure. *Marine Ecology Progress Series* **75**, 161–72.

Garrison, D.L. and Close, A.R. (1993). Winter ecology of the sea ice biota in Weddell Sea pack ice. *Marine Ecology Progress Series* **96**, 17–31.

Garrison, D.L., Sullivan, C.W. and Ackley, S.F. (1986). Sea ice microbial communities in Antarctica. *Bioscience* **36**, 243–50.

Garrison, D.L., Close, A.R. and Reinmnitz, E. (1989). Algae concentrated by frazil ice: evidence from laboratory experiments and field measurements. *Antarctic Science* **1**, 313–16.

Gendrin, A., Mangold, N., Bibring, J-P., *et al.* (2005). Sulftes in Martian layered terrains: the OMEGA/Mars Express View. *Science* **307**, 1587–91.

Gerdel, R.W. and Drouet, F. (1960). The cryoconite of the Thule area, Greenland. *Transactions of the American Microscopy Society* **79**, 256–72.

Gilbert, J.A., Davies, P.L. and Laybourn-Parry, J. (2005). A hyperactive Ca^{2+} dependent antifreeze protein in an Antarctic bacterium. *FEMS Microbiology Letters* **245**, 67–72.

Gilbert, W. (1986). The RNA world. *Nature* **319**, 618.

Glein, C.R., Zolotov, M.Y. and Shock, E.L. (2008). The oxidation state of hydrothermal systems on early Enceladus. *Icarus* **197**, 157–63.

Gleitz, M., Bartsch, A., Bieckmann, G.S. and Eicken, H. (1998). Composition and succession of sea ice diatom assemblages in the Eastern and Southern Weddell Sea, Antarctica. In Lizotte, M.P. and Arrigo, K.R. (eds) *Antarctic Sea Ice Biological Processes, Interactions and Variability*, pp 107–20. Antarctic Research Series No. 73. American Geophysical Union, Washington, DC.

Goetz, W., Bertelsen, P., Binau, C.S., *et al.* (2005). Indication of drier periods on Mars from the chemistry and mineralogy of atmospheric dust. *Nature* **436**, 62–5.

Goodchild, A., Saunders, N.F.W., Ertan, H., *et al.* (2004). A proteomic determination of cold adaptation in the Antarctic archeaon, *Methanococcides burtonii*. *Molecular Microbiology* **53**, 309–21.

Gooseff, M.N., McKnight, D.M., Lyons, W.B. and Blum, A.E. (2002). Weathering reactions and hyporheic exchange controls on stream water chemistry in a glacial meltwater stream in the McMurdo Dry Valleys. *Water Resources Research* **38**, 1279, doi 10.1029/2001WR000834.

Gooseff, M.N., McKnight, D.M., Runkel, R.L. and Duff, J.H. (2004). Denitrification and hydrologic transient storage in a glacial meltwater stream, McMurdo Dry Valleys, Antarctica. *Limnology and Oceanography* **49**, 1884–95.

Gordon, D.A., Priscu, J.C. and Giovannoni, S. (2000). Origin and phylogeny of microbes living in permanent Antarctic lake ice. *Microbial Ecology* **39**, 197–202.

Gosselin, M., Legendre, L., Therriault, J-C., De ers, S. and Rochet, M. (1986). Physical control of the horizontal patchiness of sea-ice microalgae. *Marine Ecology Progress Series* **29**, 289–98.

Gowing, M.M. (2003). Large viruses and infected microeukaryotes in Ross Sea summer pack ice habitats. *Marine Biology* **142**, 1029–40.

Gowing, M.M., Riggs, B.E., Garrison, D.L., Gibson, A.H. and Jeffries, M.O. (2002). Large viruses in Ross Sea late autumn pack ice habitats. *Marine Ecology Progress Series* **241**, 1–11.

Gowing, M.M., Garrison, D.L., Gibson, A.H., Krupp, J.M., Jeffries, M.O. and Fritsen, C.H. (2004). Bacterial and viral abundance in Ross Sea summer pack ice communities. *Marine Ecology Progress Series* **279**, 3–12.

Gradinger, R. (1999). Integrated abundance and biomass of sympagic meiofauna in Arctic and Antarctic pack ice. *Polar Biology* **22**, 169–77.

Gradinger, R. (2001). Adaptation of Arctic and Antarctic ice Metazoa to their habitat. *Zoology* **104**, 339–45.

Grannas, A.M., Jones, A.E., Dibb, J., *et al.* (2007). An overview of snow photochemistry: evidence, mechanisms and impacts. *Atmospheric Chemistry and Physics* **7**(16), 4329–73.

Grasset, Q. and Sotin, C. (1996). The cooling rate of a liquid shell in Titan's interior. *Icarus* **123**, 101–23.

Gray, D.M. and Male, D.H. (eds) (1981). *Handbook of Snow*. Pergamon Press, Ontario.

Greeley, R., Chyba, C.F., Head, J.W., *et al.* (2004). Geology of Europa. In Bagenal, F. (ed.) *Jupitor: The plants, satellites and magnetosphere*, pp 329–63. Cambridge University Press, Cambridge, UK.

Greve, R., Rupali, R.A., Mahajan, A., Segschneider, J. and Grieger, B. (2004). Evolution of the north-polar cap of Mars: a modelling study. *Planetary and Space Science* **52**, 775–87.

Gribbon, P. (1979). Cryoconite holes on Sermikavsak, West Greenland. *Journal of Glaciology* **22**(86), 177–81.

Grossart, H-P. and Simon, M. (1993). Limnetic macroscopic aggregates (lake snow): occurrence, characteristics, and microbial dynamics in Lake Constance. *Limnology and Oceanography* **38**, 532–46.

Grossmann, S. and Dieckmann, G.S. (1994). Bacterial standing stock, activity and carbon production during formation and growth of sea ice in the Weddell Sea, Antarctica. *Applied and Environmental Microbiology* **60**, 2746–53.

Guglielmo, L., Zagami, G., Saggiomo, V., Catalano, G. and Granata, A. (2007). Copepods in spring annual sea ice at Terra Nova Bay (Ross Sea, Antarctica). *Polar Biology* **30**, 747–58.

Gulley, J., Benn, D.I., Screaton, L. and Martin, J. (2009). Englacial conduit formation and implications for sub-glacial recharge. *Quaternary Science Reviews* **28**, 1984–99.

Gustafson, D.E., Stoecker, D.K., Johnson, M.D., Van Heukelem, W.F. and Sneider, K. (2000). Crytophyte algae robbed of their organelles by the marine ciliate *Mesodinium rubrum*. *Nature* **405**, 1049–52.

Haecky, P. and Andersson, A. (1999). Primary and bacterial production in sea ice in the northern Baltic Sea. *Aquatic Microbial Ecology* **20**, 107–18.

Hall, D.K., Nghiem, S.V., Schaff, C.B., *et al*. (2009). Evaluation of surface and near-surface melt characteristics on the Greenland ice sheet using MODIS and QuikSCAT data. *Journal of Geophysical Research* **114**, doi 10.1029/2009jf001287.

Hambrey, M.J. (1984). Sudden draining of ice-dammed lakes in Spitsbergen. *Polar Record* **22**, 189–94.

Hand, K.P., Chyba, C.F., Carlson, R.W. and Cooper, J.F. (2006). Clathrate hydrates of oxidants in the ice shell of Europa. *Astrobiology* **6**, 463–82.

Handelsman, J. (2004). Metagenomics: application of genomics to uncultured microorganisms. *Microbiological and Molecular Biology Reviews* **68**, 669–85.

Hanna, E., Huybrechts, P., Steffen, K., *et al*. (2008). Increased runoff from melt from the Greenland Ice Sheet: a response to global warming. *Journal of Climatology* **21**, 331–41.

Hart, J.K. and Martinez, K. (2006). Environmental sensor networks: a revolution in the earth system science? *Earth Science Reviews* **78**, 177–91.

Hart, J.K., Rose, K.C. and Martinez, K. (2011). Subglacial till behaviour derived from *in situ* wireless multi-sensor subglacial probes: rheology, hydro-mechanical interactions and till formation. *Quaternary Science Reviews* **30**, 234–47.

Hartzell, P.L., Nghiem, J.V., Richio, K.J. and Shain, D.H. (2005). Distribution and phylogeny of glacier ice worms (*Mesonchytraeus solifugus* and *Mesonchytraeus rainierensis*). *Canadian Journal of Zoology* **83**, 1206–13.

Hawes, I., Smith, R., Howard-Williams, C. and Schwarz, A.M.J. (1999). Environmental conditions during freezing, and response of microbial mats in ponds of the McMurdo Ice Shelf, Antarctica. *Antarctic Science* **11**, 198–208.

Hawes, I., Moorhead, D., Sutherland, D., Schmeling, J. and Schwarz, A-M. (2001). Benthic primary production in two perennially ice-covered Antarctic lakes, patters of biomass accumulation with a model of community metabolism. *Antarctic Science* **13**, 18–27.

Hawes, I., Howard-Williams, C.H. and Fountain, A.G. (2008). Ice-based freshwater ecosystems. In Vincent, W. and Laybourn-Parry, J. (eds) *Polar Lakes and Rivers Limnology of Arctic and Antarctic Aquatic Ecosystems*, pp 103–18. Oxford University Press, Oxford, UK.

Head, J.W. and Marchant, D.R. (2003). Cold-based mountain glaciers on Mars: Western Arsia Mons. *Geology* **31**, 641–4.

Head, J.W., Wilson, L. and Mitchell, K.L. (2003). Generation of recent massive water floods at Cerberus Fossa, Mars by dike emplacement, cryospheric cracking and confined aquifer groundwater release. *Geophysical Research Letters* **30**, doi 10.1029/2003GL017135.

Healy, F.G., Ray, R.M., Aldrich, H.C., Wilkie, A.C., Ingram, L.O. and Shanmugan, K.T. (1995). Direct isolation of functional genes encoding cellulases from the microbial consortia in a thermophilic, anaerobic digester maintained on lignocellulose. *Applied Microbiology and Biotechnology* **43**, 667–74.

Helmke, E. and Weyland, H. (1995). Bacteria in sea ice and underlying water of the eastern Weddell Sea in midwinter. *Marine Ecology Progress Series* **287**, 269–87.

Henshaw, T. and Laybourn-Parry, J. (2002). The annual patterns of photosynthesis in two large, freshwater, ultra-oligotrophic Antarctic lakes. *Polar Biology* **25**, 744–52.

Herbei, R., Lyons, W.B., Laybourn-Parry, J., Gardiner, C., Priscu, J.C. and McKnight, D.M. (2010). Physiochemical properties influencing biomass abundance and primary production in Lake Hoare, Antarctica. *Ecological Modelling* **221**, 1184–93.

Hodgkins, R., Tranter, M. and Dowdeswell, J.A. (1997). Solute provenance, transport and denudation in a high Arctic glacierized catchment. *Hydrological Processes* **11**, 1813–32.

Hodgson, D.A., Roberts, S.J., Bentley, M.J., *et al*. (2010). Exploring former subglacial Hodgson Lake. Antarctica Paper I: site description, geomorphology and limnology. *Quaternary Science Reviews* **28**, 2295–309.

Hodson, A. (2006). Biogeochemistry of snowmelt in an Antarctic glacial ecosystem. *Water Resources Research* **42**, doi 10.1029/2005WR004311.

Hodson, A., Tranter, M. and Vatne, G. (2000). Contemporary rates of chemical denundation and atmospheric CO_2 sequestration in glacier basins: an Arctic perspective. *Earth Surface Processes and Landforms* **25**, 1447–71.

Hodson, A., Mumford, P. and Lister, D. (2004). Suspended sediment and phosphorus in proglacial rivers: bioavailability and potential impacts upon the P status of ice-marginal receiving waters. *Hydrological Processes* **18**(13), 2409–22.

Hodson, A.J., Mumford, P. N., Kohler, J. and Wynn, P.M. (2005). The high Arctic glacial ecosystem: new insights from nutrient budgets. *Biogeochemistry* **72**, 233–56.

Hodson, A.J., Anesio, A.M., Ng, F., *et al*.. (2007). A glacier respires: quantifying the distribution and respiration CO_2 flux of cryoconite across an entire Arctic supraglacial ecosystem. *Journal of Geophysical Research* **112**(G04S36), doi 10.1029/2007JG000452.

Hodson, A.J., Anesio, A.M., Tranter, M., *et al*. (2008). Glacial ecosystems. *Ecological Monographs* **78**, 41–67.

Hodson, A.J., Roberts, T., Engvall, A.C., Holmén, K. and Mumford, P.N. (2009). Glacier ecosystem response to episodic nitrogen enrichment in Svalbard, European High Arctic. *Biogeochemistry* **98**, 171–84.

Hodson, A.J., Bøggild, C.E., Hanna, E., Irvine-Fynn, T., Sattler, B. and Anesio, A.M. (2010a). A microbial ecosystem upon the Greenland Ice Sheet. *Annals of Glaciology* **51**, 123–9.

Hodson, A.J., Cameron, K., Bøggild, C.E., *et al*. (2010b). The structure, biological activity and biogeochemistry of cryoconite aggregates upon an Arctic valley glacier: Longyearbyen, Svalbard. *Journal of Glaciology* **56**, 349–62.

Hoffman, P.F., Kaufman, A.J., Halverson, G.P. and Schrag, D.P. (1998). A Neoproterozoic snowball Earth. *Science* **281**, 1342–6.

Hoham, R.W. (1975). Optimum temperatures and temperature ranges for growth of snow algae. *Arctic and Alpine Research* **7**, 13–24.

Hoham, R.W. and Duval, B. (2001). Microbial ecology of snow and freshwater ice with emphasis on snow algae. In Jones, H.G., Pomeroy, J.W., Walker, D.A. and Hoham, R.W. (eds) *Snow Ecology*, pp 186–203. Cambridge University Press, Cambridge, UK.

Holland, H.D. (1978). *The Chemistry of the Atmosphere and Oceans*. Wiley, New York, 351 pp.

Holland, M.M., Bitz, C.M. and Tremblay, B. (2006). Future abrupt reductions in the summer Arctic sea ice. *Geophysical Research Letters* **33**(23), L23503.

Holmes, R.M. (2008). Lability of DOC transported by Alaskan rivers to the Arctic Ocean. *Geophyscial Research Letters* **35**, LO3402.

Holt, J.W., Safaeinili, A., Plaut, J.J., *et al*. (2008). Radar sounding evidence for buried glaciers in the southern mid-latitudes of Mars. *Science* **322**, 1235–8.

Hood, E. and Berner, L. (2009). Effects of changing glacial coverage on the physical and biogeochemical properties of coastal streams in southeastern Alaska. *Journal of Geophysical Research* **114**, doi 10.1029/2009jg000971.

Hood, E. and Scott, D. (2008). Riverine organic matter and nutrients in southeast Alaska affected by glacial coverage. *Nature Geosciences* **1**(9), 583–7.

Hood, E., Fellman, J., Spencer, R.G.M., *et al*. (2009). Glaciers as a source of ancient and labile organic matter to the marine environment. *Nature* **462**(7276), 1044–7.

Hooke, R.L. (1989). Englacial and subglacial hydrology: a qualitative review. *Arctic and Alpine Research* **21**, 221–33.

Howard-Williams, C., Priscu, J.C., and Vincent, W.F. (1989). Nitrogen dynamics in two Antarctic streams. *Hydrobiologia* **172**, 51–61.

Howard-Williams, C., Pridmore, R., Broady, P.A. and Vincent, W.F. (1990). Environmental and biological variability in the McMurdo Ice Shelf ecosystem. In Kerry, K. and Hempel, G. (eds) *Ecological Change and the Conservation of Antarctic Ecosystems*, pp 23–31. Springer-Verlag, Berlin.

Hubbard, B.P., Sharp, M.J., Willis, I.C., Nielsen, M.K. and Smart, C. (1995). Borehole water-level variations and the structure of the subglacial hydrological system of Haut Glacier d'Arolla, Valais, Switzerland. *Journal of Glaciology* **41**, 572–83.

Hudson, N., Baker, A. and Reynolds, D. (2007). Fluorescence analysis of dissolved organic matter in natural, waste and polluted waters—a review. *River Research and Applications* **23**, 631–49.

Humlum, O., Elberling, B., Hormes, A., Fjordheim, K., Hansen, O.H. and Heinemeier, J. (2005). Late-Holocene glacier growth in Svalbard, documented by subglacial relict vegetation and living soil microbes. *Holocene* **15**, 396–407.

Hyde, W.T., Crowley, T.J., Baum, S.K. and Peltier, W.R. (2000). Neoproterozic 'snowball Earth' simulations with a coupled climate/ice-sheet model. *Nature* **405**, 425–9.

Ideno, A., Yoshida, T., Iida, T., Furutani, M. and Maruyama, T. (2001). FK506-binding protein of the hyperthermophilic achaeum, *Thermococcus* sp. KS-1, a cold-shock inducible peptidyl-prolyl *cis-trans* isomerase with activities to trap and refold denatured proteins. *Biochemical Journal* **357**, 465–71.

Ikävalka, J. and Gradinger, R. (1997). Flagellates and heliozoans in the Greenland sea ice studied alive using light microscopy. *Polar Biology* **17**, 473–81.

Irvine-Fynn, T.D.L., Hodson, A.J., Moorman, B.J., Vatne, G. and Hubbard, A. (2011). Polythermal glacier hydrology: a review. *Reviews of Geophysics*. In press.

Irwin, B.D. (1990). Primary production of ice algae on a seasonally-ice-covered, continental shelf. *Polar Biology* **10**, 247–54.

James, M.R., Pridmore, R.D. and Cummings, V.J. (1995). Planktonic communities of melt pools on the McMurdo Ice Shelf, Antarctic. *Polar Biology* **15**, 555–67.

James, S.R., Burton, H.R., McMeekin, T.A. and Mancuso, C.A. (1994). Seasonal abundance of *Halomonas meridiana*, *Halomonas subglaciescola*, *Flavobacterium gondwanense* and *Flavobacterium salegens* in four Antarctic lakes. *Antarctic Science* **6**, 325–32.

Janssens, I. and Huybrechts, P. (1999). The treatment of meltwater retention in mass-balance parameterizations of the Greenland ice sheet. *Annals of Glaciology* **31**, 133–40.

Jin, M., Deal, C.J., Wang, J., *et al.* (2006). Controls of the landfast ice–ocean ecosystem offshore Barrow, Alaska. *Annals of Glaciology* **44**, 63–72.

Johnston, D.W., Freidlaende, A.S., Torres, L.G. and Lavigne, D.M. (2005). Variation in sea ice cover on the east coast of Canada from 1969 to 2002: climate variability and implications for harp and hooded seals. *Climate Research* **29**, 209–22.

Jones, H.G. (1991). Snow chemistry and biological activity: a particular perspective on nutrient cycling. In: Davies T.D, Tranter M. and Jones H.G. (eds) *Seasonal Snowpacks: Processes of compositional change*, pp 173–228. NATO ASI Series G, Ecological Sciences Vol. 28. Springer-Verlag, Berlin.

Jones, H.G. (1999). The ecology of snow-covered systems: a brief overview of nutrient cycling and life in the cold. *Hydrological Processes* **13**, 2135–47.

Jorge Villar, S.E. and Edwards, G.M. (2006). Raman spectroscopy in astrobiology. *Analytical and Bioanalytical Chemistry* **384**, 100–13.

Jouzel, J., Lorius, C., Petit, J.R., *et al.* (1987). Vostok ice core: a continuous isotope temperature record over the last climatic cycle (160,000 years). *Nature* **329**, 403–8.

Jungblut, A.D., Lovejoy, C. and Vincent, W.F. (2010). Global distribution of cyanobacterial ecotypes in the cold biosphere. *ISME Journal* **4**, 191–202.

Junge, K., Imhoff, F., Staley, T. and Deming, J.W. (2002). Phylogenetic diversity of numerically important Arctic sea-ice bacteria cultured at subzero temperature. *Microbial Ecology* **43**, 315–28.

Kaartokallio, H. (2004). Food web components and physical and chemical properties of Baltic sea ice. *Marine Ecology Progress Series* **273**, 49–63.

Kamb, B. (1987). Glacier surge mechanism based on linked cavity configuration of the basal water conduit system. *Journal of Geophysical Research—Solid Earth and Planets* **92**, 9083–100.

Kamb, B., Raymond, C.F., Harrison, W.D., *et al.* (1985). Glacier surge mechanism:1982–1983 surge of Variegated Glacier, Alaska. *Science* **227**, 469–79.

Kang, S-H. and Fryxell, G.A. (1992). *Fragilariopsis cylindrus* (Grunow) Kreiger: the most abundant diatom in water column assemblages of Antarctic marginal ice-edge zones. *Polar Biology* **12**, 609–27.

Kapitsa, A.P., Ridley, J.K., Robin, G.D., Siegert, M.J. and Zotikov, I.A. (1996). A large deep freshwater lake beneath the ice of central East Antarctica. *Nature* **381**, 684–6.

Kargel, J.S. (1998). The salt of Europa. *Science* **280**, 1211–12.

Kargel, J.S. and Strom, R.G. (1991). Ancient glaciation on Mars. *Geology* **20**, 3–7.

Kargel, J.S., Kaye, J.Z., Head, J.W., *et al.* (2000). Europa's crust and ocean: origin, composition, and the prospects for life. *Icarus* **148**, 226–65.

Karl, D.M., Bird, D.F., Björkman, K., Houlihan, T., Shackelford, R. and Tupas, L. (1999). Microorganisms in the accreted ice of Lake Vostok, Antarctica. *Science* **286**, 2144–7.

Kelley, J.J., Weaver, D.F. and Smith, B.P. (1968). The variation of carbon dioxide under snow in the Arctic. *Ecology* **49**, 358–61.

Kennedy, M.J., Runnegar, B., Prave, A.R., Hoffmann, K-H. and Arthur, M.A. (1998). Two or four Neoproterozoic glaciations? *Geology* **26**, 1059–63.

Kern, J.C. and Carey, A.G. (1983). The faunal assemblage inhabiting seasonal sea ice in the nearshore Arctic Ocean with emphasis on copepods. *Marine Ecology Progress Series* **10**, 159–67.

Kieffer, H.H. and Zent, A.P. (1992). Quasi-periodic climate change on Mars. In Kieffer, H.H., Jakosky, B.M., Snyder, C.W. and Matthews, M.S. (eds) *Mars*, pp 1180–218. University of Arizona Press, Tucson, AZ.

Kieffer, H.H., Christensen, P.R. and Titus, T.T. (2006). CO_2 jets formed by sublimation beneath translucent slab ice in Mars' seasonal south polar ice cap. *Nature* **442**, 793–6.

Kiko, R., Michels, J., Mizdalski, E., Schnack-Schiel, S.B. and Werner, I. (2008). Living conditions, abundance and composition of the metazoan fauna in surface and sub-ice layers in pack ice of the western Weddell Sea during late spring. *Deep Sea Research II* **55**, 1000–14.

King, E.C., Smith, A.M., Murray, T. and Stuart, G.W. (2008). Glacier bed characteristics of Midtre Lovenbreen, Svalbard, from high resolution seismic and radar surveying. *Journal of Glaciology* **54**, 145–57.

Konovalov, V. (2000). Computations of melting under moraine as a part of a regional modelling of glacier runoff. In Nakawo, M., Raymond, C.F. and Fountain, A (eds) *Symposium at Seattle 2000 on Debris-Covered Glaciers*, pp 109–18. International Association of Hydrological Sciences Publication No. 264, IAHS Press, Oxfordshire, UK.

Koonin, E.V. and Wolf, Y.I. (2008). Genomics of bacteria and archaea: emerging dynamic view of the prokaryote world. *Nucleic Acids Research* **36**, 6688–719.

Kottmeier, S.T. and Sullivan, C.W. (1987). Late winter primary production and bacterial production in sea ice and seawater west of the Antarctic Peninsula. *Marine Ecology Progress Series* **36**, 287–98.

Krembs, C. and Engel, A. (2001). Abundance and variability of microorganisms and transparent exopolymer particles across the ice–water interface of melting first-year ice in the Laptev Sea (Arctic). *Marine Biology* **138**, 173–85.

Krembs, C., Gradinger, R. and Spindler, M. (2000). Implications of brine channel geometry and surface area for the interaction of sympagic organisms in Arctic sea ice. *Journal of Experimental Marine Biology and Ecology* **243**, 55–80.

Kuhn, M. (2001). The nutrient cycle through snow and ice, a review. *Aquatic Sciences* **63**, 150–67.

Kumar, A., Perlwitz, J., Eischeid, J., *et al.* (2010). Contribution of sea ice loss to Arctic amplification. *Geophysical Research Letters* **37**(21), L21701.

Kumar, G.S., Jagannadham, M.V. and Ray, M.K. (2002). Low-temperature-induced changes in composition and fluidity of lipopolysaccharides in the Antarctic psychrotrophic bacterium *Pseudomonas syringae*. *Journal of Bacteriology* **184**, 6746–9.

Kwok, R. and Morison, J. (2011). Dynamic topography of the ice-covered Arctic Ocean from ICESat. *Geophysical Research Letters* **38**(2), L02501.

Kwok, R., Cunningham, G.F., Wensnahan, M., Rigor, I., Zwally, H.J. and Yi, D. (2009). Thinning and volume loss of the Arctic Ocean sea ice cover: 2003–2008. *Journal of Geophysical Research* **114**(C07005), doi 10.1029/2009JC 005312.

Lafreniére, M.J. and Sharp, M.J. (2004). The concentration and fluorescence of dissolved organic carbon (DOC) in glacial and nonglacial catchments: interpreting hydrological flow routing and DOC sources. *Arctic, Antarctic and Alpine Research* **36**, 156–65.

Laj, P., Palais, J.M. and Sigurdsson, H. (1992). Changing sources of impurities to the Greenland ice-sheet over the last 250 years. *Atmospheric Environment* **26**, 2627–40.

Lane, D.J., Pace, B., Olsen, G.J., Stahl, D.A., Sogin, M.L. and Pace, N.R. (1985). Rapid determination of 16S ribosomal RNA sequences for phylogenetic analysis. *Proceedings of the National Academy of Sciences* **82**, 6955–9.

Langevin, Y., Poulet, F., Bibring, J.P. and Gondet, B. (2005). Sulphates in the north polar region of Mars detected by OMEGA/Mars Express. *Science* **307**, 1584–6.

Langford, H., Hodson, A.J., Banwart, S.A. and Bøggild, C.E. (2010). The microstructure and biogeochemistry of Arctic cryoconite granules, *Annals of Glaciology* **51**, 87–91.

Langmuir, D. (1997). *Aqueous Environmental Geochemistry*. Prentice-Hall, Englewood Cliffs, NJ, 600 pp.

Lanoil, B., Skidmore, M., Priscu, J.C., *et al.* (2009). Bacteria beneath the West Antarctic ice sheet. *Environmental Microbiology* **11**, 609–15.

Laurion, I., Demers, S. and Vézina, A.F. (1995). The microbial food web associated with the ice algal assemblage: biomass and bactivory of nanoflagellate protozoans in Resolute Passage (High Canadian Arctic). *Marine Ecology Progress Series* **120**, 77–87.

Laybourn-Parry, J. (1984). *A Functional Biology of Free-living Protozoa*. Croom-Helm, London.

Laybourn-Parry, J. (2009). No place too cold. *Science* **324**, 1521–2.

Laybourn-Parry, J. and Marshall, W.A. (2003). Photosynthesis, mixotrophy and microbial plankton dynamics in two high Arctic lakes during summer. *Polar Biology* **26**, 517–24.

Laybourn-Parry, J. and Vincent, W.F. (2008). Future directions. In Vincent, W.F. and Laybourn-Parry, J. (eds) *Polar Lakes and Rivers: Limnology of Arctic and Antarctic aquatic ecosystems*, pp 307–16. Oxford University Press, Oxford, UK.

Laybourn-Parry, J., Henshaw, T. and Quayle, W.C. (2002). The evolution and biology of Antarctic saline lakes in relation to salinity and trophy. *Polar Biology* **25**, 542–52.

Laybourn-Parry, J., Henshaw, T., Jones, D.J. and Quayle, W. (2004). Bacterioplankton production in freshwater Antarctic lakes. *Freshwater Biology* **49**, 735–44.

Laybourn-Parry, J., Marshall, W.A. and Marchant, H.J. (2005). Flagellate nutritional versatility as a key to survival in two contrasting Antarctic saline lakes. *Freshwater Biology* **50**, 830–8.

Laybourn-Parry, J., Madan, N.J., Marshall, W.A., Marchant, H.J. and Wright, S.W. (2006). Carbon dynamics in an ultra-oligotrophic epishelf lake (Beaver Lake, Antarctica) in summer. *Freshwater Biology* **51**, 1116–30.

Leadbeater, B.S.C. and Green, J.C. (eds) (2000). *The Flagellates: Unity, diversity and evolution*. Taylor and Francis, London.

Legrand, M.R. and Delmas, R.J. (1984). The ionic balance of Antarctic snow—a 10-year detailed record. *Atmospheric Environment* **18**, 1867–74.

Leppäranta, M. and Wang, K. (2008). The ice cover on small and large lakes: scaling analysis and mathematical modeling. *Hydrobiologia* **599**, 183–9.

Leu, E., Wiktor, J., Søreide, J.E., Berge, J. and Falk-Petersen, S. (2010). Increased irradiance reduces food quality of sea ice algae. *Marine Ecology Progress Series* **411**, 49–60.

Light, J.J. and Belcher, J.H. (1968). A snow microflora in the Cairngorm Mountains, Scotland. *British Phycology Bulletin* **3**, 471–3.

Lin, L.H., Wang, P.L., Rumble, D., *et al.* (2006). Long-term sustainability of a high-energy, low-diversity crustal biome. *Science* **314**, 479–82.

Lindholm, T. (1985). *Mesodinium rubrum*—a unique photosynthetic ciliate. *Advances in Aquatic Microbiology* **3**, 1–48.

Lipps, J.H. and Rieboldt, S. (2005). Habitats and taphonomy of Europa. *Icarus* **177**, 515–27.

Lipson, D.A., Monson, R.K., Schmidt, S.K. and Weintraub, M.N. (2009a). The trade-off between growth rates and yield in microbial communities and the consequences for under-snow soil respiration in a high elevation coniferous forest. *Biogeochemistry* **95**, 23–35.

Lipson, D.A., Williams, M.W., Helmig, D., *et al.* (2009b). Process-level controls on CO_2 from a seasonally snow-covered subalpine meadow soil, Niwot Ridge, Colorado. *Biogeochemistry* **95**, 151–66.

Liston, G.E. and Winther, J-G. (2005). Antarctic surface and subsurface snow and ice melt fluxes. *Journal of Climatology* **18**, 1469–81.

Lizotte, M.P. (2001). The contribution of sea ice algae to Antarctic primary production. *American Zoologist* **41**, 51–73.

Lizotte, M.P. and Sullivan, C.W. (1992). Photosynthetic capacity in microalgae associated with Antarctic pack ice. *Polar Biology* **12**, 497–502.

Lliboutry, L. (1996). Temperate ice permeability, stability of water veins and percolation of internal meltwater. *Journal of Glaciology* **42**, 201–11.

Lyons, W.B., Wake, C. and Mayewski, P.A. (1991). Chemistry of snow at high altitude, mid/low latitude glaciers. In Davies, T.D., Tranter, M. and Jones, H.G. (eds) *Seasonal Snowpacks: Processes of compositional change*, pp 359–84. NATO ASI Series G, Ecological Sciences Vol. 28. Springer-Verlag, Berlin.

Lyons, W.B., Welch, K.A., Snyder, G., *et al.* (2005). Halogen geochemistry of the McMurdo dry valley lakes, Antarctica: clues to the origin of solutes and lake evolution. *Geochimica et Cosmochimica Acta* **69**, 305–23.

Madan, N.J., Marshall, W.A. and Laybourn-Parry, J. (2005). Virus and microbial loop dynamics over an annual cycle in three contrasting Antarctic lakes. *Freshwater Biology* **50**, 1291–1300.

Mader, H., Pettitt, M.E., Wadham, J.L., Wolff, E. and Parkes, R.J. (2006). Subsurface ice as a microbial habitat. *Geology* **34**, 169–72.

Magnusen, J.J., Robertson, D.M., Benson, B.J., *et al.* (2000). Historical trends in lake and river ice cover in the Northern Hemisphere. *Science* **289**, 1743–6.

Mangold, N., Quantin, C., Ansan, V., Delacourt, C. and Allemand, P. (2004). Evidence for precipitation on Mars from dendritic valleys in the Valles Marineris area. *Science* **305**, 781.

Manning, C.V., Zahnle, K.J. and McKay, C.P. (2009). Impact processing of nitrogen on early Mars. *Icarus* **199**, 273–85.

Maranger, R. and Bird, D.F. (1995). Viral abundance in aquatic systems: a comparison between marine and fresh waters. *Marine Ecology Progress Series* **121**, 217–26.

Maranger, R., Bird, D.F. and Juniper, S.K. (1994). Viral and bacterial dynamics in Arctic sea ice during the spring algal bloom near Resolute, NWT, Canada. *Marine Ecology Progress Series* **111**, 121–7.

Marchant, H.J. and Scott, F.J. (1993). Uptake of submicrometer particles and dissolved organic matter by Antarctic choanoflagellates. *Marine Ecology Progress Series* **92**, 59–64.

Marchant, H., Davidon, A., Wright, S. and Glazebrook, J. (2000). The distribution and abuandance of viruses in the Southern Ocean during spring. *Antarctic Science* **12**, 414–17.

Margulis, L. (1974). Five Kingdom classification and origin and evolution of cell. In Dobzansky, T., Hecht, M.K. and Steere, W.C. (eds) *Evolutionary Biology*, Vol. 7, pp 45–78. Plenum Press, New York.

Marion, G.M., Fritsen, C.H., Eicken, H. and Payne, M.C. (2003). The search for life on Europa: limiting environmental factors, potential habitats, and Earth analogue. *Astrobiology* **3**, 785–811.

Markager, S., Vincent, W.F. and Tang, E.P.Y. (1999). Carbon fixation in high Arctic lakes: implications of low temperature for photosynthesis. *Limnology and Oceanography* **44**, 597–607.

Markus, T., Stroeve, J.C. and Miller, J. (2009). Recent changes in Arctic sea ice melt onset, freezeup, and melt season length. *Journal of Geophysical Research* **114**(C12024), doi 10.1029/2009JC005436.

Marsh, A.R.W. and Webb, A.H. (1979). *Physico-chemical Aspects of Snow-melt*. Central Electricity Generating Board Reports RD/L/N 60/79. Central Electricity Generating Board, London, 12 pp.

Marsh, P. and Woo, M.K. (1984). Wetting front advance and freezing of meltwater within a snow cover: 2. A simulation model. *Water Resources Research* **20**, 1865–74.

Marsh, P. and Woo, M.K. (1985). Meltwater movement in natural heterogeneous snow covers. *Water Resources Research* **21**, 1710–16.

Matson, D.L., Castillo, J.C., Lunine, J. and Johnson, T.V. (2007). Enceladus's plume: compositional evidence for a hot interior. *Icarus* **187**, 569–73.

Mätzler, C. (1998). Microwave properties of ice and snow. In: Schmitt, B., de Bergh, C. and M. Festou (eds) *Solar System Ices*, pp 241–57. Kluwer Academic Publications, Dordrecht.

McIntyre, N.F. (1984). Cryoconite hole thermodynamics. *Canadian Journal of Earth Sciences* **21**(2), 152–6.

McKay, C.P. (2000). Thickness of tropical ice and photosynthesis on a snowball Earth. *Geophysical Research Letters* **27**, 2153–6.

McKay, C.P. and Smith, H.D. (2005). Possibilities for methanogenic life in liquid methane on the surface of Titan. *Icarus* **178**, 274–6.

McKay, C.P., Hand, K.P., Doran, P.T., Andserson, D.T. and Priscu, J.C. (2003). Clathrate formation and the fate of noble and biologically useful gases in Lake Vostok. *Geophysical Research Letters* **30**, doi 10.1029/2003GL 017490,35-1-35-4.

Mckay, C.P., Porco, C.C., Altheide, T., Davis, W.L. and Kral, T.A. (2008). The possible origin and persistence of life on Enceladus and detection of biomarkers in the plume. *Astrobiology* **8**, 909–19.

McKay, D.S., Everett, K.G., Thomas-Keprta, K.L., *et al.* (1996). Search for past life on Mars: possible relic biogenic activity in Martian meteorite ALH84001. *Science* **273**, 924–30.

McKinnon, W.B. and Zolensky, M.E. (2003). Sulphate content of Europa's ocean and shell: evolutionary considerations and some geological and astrobiological implications. *Astrobiology* **3**, 879–97.

McMillan, M., Nienow, P., Shepherd, A., Benham, T. and Sole, A. (2008). Seasonal evolution of supra-glacial lakes on the Greenland Ice Sheet. *Earth and Planetary Science Letters* **262**, 484–92.

McMinn, A. and Hegseth, E.N. (2003). Early spring pack ice algae from the Arctic and Antarctic: how different are they? In Huiskies, A.H.L., Gieskes, W.W.C., Rozema, J., Schorno, R.M.L., van der Vies, S.M. and Wolff, W.J. (eds) *Antarctic Biology in a Global Context*, pp 182–6. Backhuys Pulishers, Leiden, Netherlands.

McMinn, A. and Hegseth, E.N. (2007). Sea ice primary production in the northern Barents Sea, spring 2004. *Polar Biology* **30**, 289–94.

McMinn, A., Ryan, K. and Gademann, R. (2003). Diurnal changes in photosynthesis of Antarctic fast ice algal communities determined by pulse amplitude modulated fluorimetry. *Marine Biology* **143**, 359–67.

McMinn, A., Ryan, K.G., Ralph, P.J. and Pankowski, A. (2007). Spring sea ice photosynthesis, primary productivity and biomass distribution in eastern Antarctica, 2002–2004. *Marine Biology* **151**, 985–95.

Meiners, K., Fehling, J., Granskog, M.A. and Spindler, M. (2002). Abundance, biomass and composition of biota in Baltic Sea ice and underlying water. *Polar Biology* **25**, 761–70.

Meiners, K., Brinkmeyer, R., Granskog, M.A. and Lindfors, A. (2004). Abundance, size distribution and bacterial colonization of exoploymer particles in Antarctic sea ice (Bellinghausen Sea). *Aquatic Microbial Ecology* **35**, 283–96.

Melnikov, I.A. (2005). Sea ice–upper ocean ecosystems and global changes in the Arctic. *Russian Journal of Marine Biology* **31**, S1–S8.

MEPAG (Mars Exploration Analysis Group) (2006). Findings of the Mars Special Regions Science Analysis Group. *Astrobiology* **6**, 677–732.

Mernild, S.H., Liston, G.E., Steffen, K., van den Broeke, M. and Hasholt, B. (2010). Runoff and mass-balance simulations from the Greenland Ice Sheet at Kangerlussuaq (Søndre Strømfjord) in a 30-year perspective, 1979–2008. *Cryosphere* **4**(2), 231–42.

Metz, J.G., Roessier, P., Facciotti, D., *et al.* (2001). Production of polyunsaturated fatty acids by polyketide synthases in both prokaryotes and eukaryotes. *Science* **292**, 290–3.

Meyer, J.L., Edwards, R.T. and Risley, R. (1997). Bacterial growth on dissolved organic carbon from a blackwater river. *Microbial Ecology* **13**, 13–29.

Michel, C., Legendre, L., Demers, S. and Therriault, J.C. (1988). Photoadaptation of sea-ice microalgae in springtime: photosynthesis and carboxylating enzymes. *Marine Ecology Progress Series* **50**, 177–85.

Mikucki, J. and Priscu, J.C. (2007). Bacterial diversity associated with Blood Falls, a subglacial outflow from the Taylor Glacier, Antarctica. *Applied and Environmental Microbiology* **73**, 4029–39.

Mikucki, J., Foreman, C.M., Sattler, B., Lyons, W.B. and Priscu, J.C. (2004). Geomicrobiology of Blood Falls: an iron-rich saline discharge at the terminus of the Taylor Glacier, Antarctica. *Aquatic Geochemistry* **10**, 199–220.

Mikucki, J., Pearson, A., Johnston, D.T., *et al.* (2009). A contemporary microbially maintained subglacial ferrous 'ocean'. *Science* **324**, 397–9.

Mindl, B., Anesio, A.M., Meirer, K., *et al.* (2007). Factors influencing bacterial dynamics along a transect from supraglacial runoff to proglacial lakes of a high Arctic glacier. *FEMS Microbiology Ecology* **59**, 307–17.

Miteva, V.I. (2007). Bacteria in snow and glacier ice. In Margesin, R., Schinner, F., Marx, J-C. and Gerday, C. (eds) *Psychrophiles: From biodiversity to biotechnology*, pp 31–50. Springer-Verlag, Berlin.

Miteva, V.I. and Brenchley, J.E. (2005). Detection and isolation of ultrasmall microorganisms from a 120,000-year-old Greenland ice core. *Applied and Environmental Microbiology* **71**, 7806–18.

Miteva, V.I., Sheridan, P.P. and Brenchley, J.E. (2004). Phylogenetic and physiological diversity of microorganisms isolated from a deep Greenland glacier ice core. *Applied and Environmental Microbiology* **70**, 202–13.

Miteva, V.I., Teacher, C., Sowers, T. and Brenchley, J.E. (2009). Comparison of the microbial diversity at different depths of the GISP2 Greenland ice core in relationship to deposition climates. *Environmental Microbiology* **11**, 640–56.

Mock, T. and Thomas, D.N. (2005). Recent advances in sea-ice microbiology. *Environmental Microbiology* **7**, 605–19.

Mondino, L.J., Asao, M. and Madigan, M.T. (2009). Cold-active bacteria from ice-sealed Lake Vida, Antarctica. *Achives of Microbiology* **191**, 785–90.

Monfort, P. and Baleux, B. (1992). Comparison of flow cytometry and epifluorescene microscopy for counting bacteria in aquatic ecosystems. *Cytometry* **13**, 188–92.

Monson, R.K., Lipson, D.L., Burns, S.P., *et al.* (2006). Winter forest soil respiration by climate and microbial community composition. *Nature* **439**, 711–14.

Moorhead, D.L., Davis, W.S. and Wharton, R.A. (1997). Carbon dynamics of aquatic microbial mats in the Antarctic Dry Valleys: a modeling synthesis. In Lyons, W.B., Howard-Williams, C. and Hawes, I. (eds) *Ecosystem Processes in Antarctic Ice-free Landscapes*, pp 181–96. A.A. Balkema, Rotterdam.

Moorhead, D.L., Schmeling, J. and Hawes, I. (2005). Modeling the contribution of benthic microbial mats to net primary production in Lake Hoare, McMurdo Dry Valleys. *Antarctic Science* **17**, 33–45.

Moorthi, S., Caron, D.A., Gast, R.J. and Sanders, R.W. (2009). Mixotrophy: a widespread and important ecological strategy for planktonic and sea-ice nanoflagellates in the Ross Sea, Antarctica. *Aquatic Microbial Ecology* **54**, 269–77.

Moseley-Thompson, E., Paskievitch, J.F., Gow, A.J. and Thompson, L.G. (1999). Late 20th century increase in South Pole accumulation. *Journal of Geophysical Research* **104**(D4), 3877–86.

Mosier, A.C., Murray, A.E. and Fritsen, C.H. (2007). Microbiota within the perennial ice cover of Lake Vida, Antarctica. *FEMS Microbiology Ecology* **59**, 274–88.

Mueller, D.R., Vincent, W.F., Pollard, W.H. and Fritsen, C.H. (2001). Glacial cryoconite ecosystems: a bipolar comparison of algal communities and habitats. *Nova Hedwigia* **123**, 173–97.

Mueller, D.R., Vincent, W.F., Bonilla, S. and Laurion, I. (2005). Extremophiles and broadband pigmentation strategies in a high arctic ice shelf ecosystem. *FEMS Microbiology Ecology* **53**, 73–87.

Mueller, D.R., Vincent, W.F. and Jeffries, M.O. (2006). Environmental gradients, fragmented habitats, and microbiota of a northern ice shelf cryoecosystem, Ellesmere Island, Canada. *Arctic, Antarctic and Alpine Research* **38**, 593–607.

Müller, F. (1961). Zonation in the accumulation area of the glaciers of Axel Heiberg Island, N.W.T., Canada. *Journal of Glaciology* **4**(33), 302–210.

Müller, F. and Keeler, C.M. (1969). Errors in short-term ablation measurements on melting ice surfaces. *Journal of Glaciology* **8**, 91–105.

Murray, J.B., Muller, J-P., Neukum, G., *et al.* (2005). Evidence from Mars Express High Resolution Stereo Camera from a frozen sea close Mars equator. *Nature* **434**, 352–5.

Nealson, K.H., Tsapin, A. and Storrie-Lombardi, M. (2002). Searching for life in the universe: unconventional methods for an unconventional problem. *International Microbiology* **5**, 223–30.

Notz, D. (2009). The future of ice sheets and sea ice: between reversible retreat and unstoppable loss. *Proceedings of the National Academy of Sciences* **106**(49), 20590–5.

Novak, R.E., Mumma, M.J. and Villanueva, G.L. (2011). Measurement of the isotopic signatures of water on Mars; implications for studying methane. *Planetary and Space Science* **59**, 163–8.

Ohno, H., Igarashi, M. and Hondoh, T. (2005). Salt inclusions in polar ice core: location and chemical form of water-soluble impurities. *Earth and Planetary Science Letters* **232**, 171–8.

Oswald, G.K.A. and Gogieni, S.P. (2008). Recovery of subglacial water extent from Greenland radar survey data. *Journal of Glaciology* **54**(184), 94–106.

Pace, N.R., Stahl, D.A., Lane, D.J. and Olsen, G.J. (1985). Analyzing natural microbial populations by rRNA sequences. *American Society for Microbiology News* **51**, 4–12.

Paerl, H.W. and Priscu, J.C. (1998). Microbial phototrophic, heterotrophic, and diazotrophic activities associated with aggregates in the permanent ice cover of Lake Bonney, Antarctica. *Microbial Ecology* **36**(3), 221–30.

Painter, T.H., Duval, B., Thomas, W.H., Mendez, M., Heintzelman, S. and Dozier, J. (2001). Detection and quantification of snow algae with an airborne imagining spectrophotometer. *Applied and Environmental Microbiology* **67**, 5267–72.

Palmisano, A.C. and Sullivan, C.W. (1983). Sea ice microbial communities (SIMCO). 1. Distribution, abundance and primary production of ice microalgae in McMurdo Sound, Antarctica in 1980. *Polar Biology* **2**, 171–7.

Palmisano, A.C., Beeler SooHoo, J. and Sullivan, C.W. (1987). Effect of environmental variables on photosynthesis–irradiance relationships in Antarctic sea-ice microalgae. *Marine Biology* **94**, 299–306.

Parker, T.L., Saunders, R.S. and Schneebergwr, D.M. (1989). Transitional morphology in West Deuteronilus Mensae, Mars: implications for modifications of the lowland/upland boundary. *Icarus* **82**, 111–45.

Parkinson, C.D., Liang, M.C., Hartman, H., *et al.* (2007). Enceladus: Cassini observations and implications for the search for life. *Astronomy and Astrophysics* **463**, 353–7.

Paterson, H. and Laybourn-Parry, J. Sea-ice microbial dynamics over an annual ice cycle. Submitted to *Polar Biology*.

Paterson, H. and Laybourn-Parry, J. (2011) Viral dynamics in the fast ice of Prydz Bay, Antarctica. *Polar Biology*. In press, doi: 10.1007/s00300-011-1093-z.

Paterson, W.S.B. (1994). *The Physics of Glaciers*, 3rd edn. Pergamon Press, Oxford, UK.

Patterson, D.J. and Hausmann, K. (1981). Feeding by *Actinophrys sol* (Protozoa: Heliozoa) 1. Light microscopy. *Microbios* **31**, 39–55.

Pauer, F., Kipfstuhl, J. and Kuhs, W.F. (1997). Raman spectroscopic and statistical studies on natural clathrates from the Greenland ice core, and neutron diffraction studies on synthetic nitrogen clathrates. *Journal of Geophysical Research C Oceans* **102**, 519–26.

Pearce, D.A. (2003). Bacterioplankton community structure in a maritime Antarctic oligotrophic lake during a period of holomixis, as determined by denaturing gradient gel electrophoresis (DGGE) and fluorescence in situ hybridisation (FISH). *Microbial Ecology* **46**, 92–105.

Pearce, D.A., Van der Gast, C.J., Lawley, B. and Ellis-Evans, J.C. (2003). Bacterioplankton community diversity in a maritime Antarctic lake, determined by culture-dependent and culture-independent techniques. *FEMS Microbiology Ecology* **45**, 59–70.

Pearce, D.A., Van der Gast, C.J., Woodward, K. and Newsham, K.K. (2005). Significant changes in the bacterioplankton community structure of a maritime Antarctic freshwater lake following nutrient enrichment. *Microbiology SGM* **151**, 3237–48.

Pearce, I., Davidson, A.T., Bell, E.M. and Wright, S. (2007). Seasonal changes in the concentration and metabolic activity of bacteria and viruses at an Antarctic coastal site. *Aquatic Microbial Ecology* **47**, 11–23.

Pellerin, B.A., Downing, B.D., Kendall, C., *et al.* (2009). Assessing the sources and magnitude of diurnal nitrate variability in the San Joaquin River (California) with an *in situ* optical nitrate sensor and dual nitrate isotopes. *Freshwater Biology* **54**, 376–87.

Perriss, S.J. and Laybourn-Parry, J. (1997). Microbial communities in the saline lakes of the Vestfold Hills (Eastern Antarctica). *Polar Biology* **18**, 135–44.

Perriss, S.J., Laybourn-Parry, J. and Marchant, H.J. (1995). The widespread occurrence of the autotrophic ciliate *Mesodinium rubrum* (Ciliophora: Haptorida) in the freshwater and brackish lakes of the Vestfold Hills, Eastern Antarctica. *Polar Biology* **15**, 423–8.

Perry, J.J., Staley, J.T. and Lory, S. (2002). *Microbial Life*. Sinauer Associates, Sunderland, MA.

Peterson, B.J., McClelland, J., Curry, R., Holmes, R.M., Walsh, J.E. and Aagaard, K. (2006). Trajectory shifts in the Arctic and subarctic freshwater cycle. *Science* **313**(5790), 1061–6.

Petit, J.R., Jouzel, J., Raynaud, D., *et al.* (1999). Climate and atmospheric history of the past 420,000 years from the Vostok ice core, Antarctica. *Nature* **399**, 429–36.

Petters, M.D., Parsons, M.T., Prenni, A.J., *et al.* (2009). Ice nuclei emissions from biomass burning. *Journal of Geophysical Research* **114** (D07209), doi 10.1029/2008JD011532.

Pettersson, R., Jansson, P. and Blatter, H. (2004). Spatial variability in water content at the cold temperate transition surface of the polythermal Storglaciaren, Sweden. *Journal of Geophysical Research* **109**(F02009).

Phillips, R.J., Zuber, M.T., Smrekar, S.E., *et al.* (2008). Mars north polar deposits: stratigraphy, age, and geodynamical response. *Science* **320**, 1182–5.

Pierazzo, E. and Chyba, C.F. (1999). Amino acid survival in large cometary impacts. *Meteoritic, Planetary and Space Science* **34**, 909–18.

Pierazzo, E. and Chyba, C.F. (2002). Cometary delivery of biogenic elements to Europa. *Icarus* **157**, 120–27.

Piquet, A.M.T., Bolhuis, H., Meredith, M.P. and Buma, A.G.J. (2011). Shifts in coastal Antarctic marine microbial communities during and after melt water related surface stratification. *FEMS Microbiology Ecology* **76**, 413–27.

Plaut, J.J., Picardi, G., Safaeinili, A., *et al.* (2007). Subsurface radar sounding of the south polar layered deposits of Mars. *Science* **316**, 92–5.

Pogson, L., Tremblay, B., Lavoie, D. and Vancoppenolle, M. (2011). Development and validation of a one-dimensional snow-ice algae model against observations in Resolute Passage, Canadian Arctic Archipelago. *Journal of Geophysical Research* **116**(C04010), doi 10.1029/2010JC 006119.

Pomeroy, L.R. (1974). The ocean's food web, a changing paradigm. *Bioscience* **9**, 499–504.

Pondrelli, M., Pio Rossi, A., Marinangeli, L., *et al.* (2008). Evolution and depositional environments of the Eberswalde fan delta, Mars. *Icarus* **197**, 429–51.

Porco, C.C., Helfenstein, P., Thomas, P.C., *et al.* (2006). Cassini observes the active south pole of Enceladus. *Science* **311**, 1393–401.

Porazinska, D.L., Foutian, A.G., Nylen, T.H., Tranter, M., Virginia, R.A. and Wall, D.H. (2004). The biodiversity and biogeochemistry of cryoconite holes from McMurdo Dry Valley glaciers, Antarctica. *Arctic, Antarctic and Alpine Research* **36**, 84–91.

Powner, M.W., Gerland, B. and Sutherland, J.D. (2009). Synthesis of activated pyrimidine ribonucleotides in prebiotically plausible conditions. *Nature* **459**, 239–42.

Price, P.B. (2000). A habitat for psychrophiles in deep, Antarctic ice. *Proceedings of the National Academy of Science USA* **97**, 1247–51.

Price, B.P. (2007). Microbial life in glacial ice and implications for a cold origin of life. *FEMS Microbial Ecology* **59**, 217–31.

Price, B.P. (2009). Microbial genesis, life and death in glacial ice. *Canadian Journal of Microbiology* **55**, 1–11.

Price, B.P., Rohde, R.A. and Bay, R.C. (2009). Flux of microbes, organic aerosols, dust, sea-salt Na ions, non sea-salt Ca ions and methanosulfonate onto Greenland and Antarctic ice. *Biogeosciences* **6**, 479–86.

Prieto-Ballesteros, O., Rodriguez, N., Kargel, J.S., Kessler, C.G., Amils, R. and Remolar, D.F. (2003). Tirez Lake as a terrestrial analogue of Europa. *Astrobiology* **3**, 863–77.

Priscu, J.C., Fritsen, C.H., Adams, E.E., *et al.* (1998). Perennial Antarctic lake ice: an oasis for life in a polar desert. *Science* **280**, 2095–8.

Priscu, J.C., Adams, E.E., Lyons, W.B., *et al.* (1999). Geomicrobiology of subglacial ice above Lake Vostok, Antarctica. *Science* **286**, 2141–4.

Priscu, J.C., Adams, E.E., Paerl, H.W., *et al.* (2005). Perennial Antarctic ice: a refuge for cyanobacteria in an extreme environment. In Castello, J.D. and Rogers, S.O. (eds) *Life in Ancient Ice*, pp 22–49. Princeton University Press, Princeton, NJ.

Priscu, J.C., Christner, B.C., Foreman, C.M. and Royston-Bishop, G. (2006). Biological material in ice cores. In Elias, S. (ed.) *Encyclopedia of Quaternary Sciences*, pp 1156–66. Elsevier, London.

Priscu, J.C., Tulczyk, S., Studinger, M., Kennicutt, M.C., Christner, B.C. and Forman, C.M. (2008). Antarctic subglacial water: origin, evolution and ecology. In: Vincent, W.F. and Laybourn-Parry, J. (eds) *Polar Lakes and Rivers: Limnology of Arctic and Antarctic aquatic systems*, pp 119–36. Oxford University Press, Oxford, UK.

Psenner, R. and Sattler, B. (1998). Life at freezing point. *Science* **280**, 2073–4.

Puxbaum, H., Kovar, A. and Kalina, M. (1991). Chemical composition and fluxes at elevated sites (700–3105 m asl) in the eastern Alps (Austria). In Davies, T.D., Tranter, M. and Jones, H.G. (eds) *Seasonal Snowpacks: Processes of compositional change*, pp 273–98. NATO ASI Series G, Ecological Sciences Vol. 28. Springer-Verlag, Berlin.

Quetin, L.B. and Ross, R.M. (2003). Episodic recruitment in Antarctic krill. *Marine Ecology Progress Series* **259**, 185–200.

Quetin, L.B., Ross, R.M., Frazer, T.K., Amsler, M.O., Wyatt-Evans, C and Oakes, S.A. (2003). Growth of larval krill, *Euphausia superba*, in fall and winter west of the Antarctic Peninsula. *Marine Biology* **143**, 833–43.

Raiswell, R., Benning, L.G., Davidson, L., Tranter, M. and Tulaczyk, S. (2009). Schwertmannite in wet, acid and oxic microenvironments beneath polar and polythermal glaciers. *Geology* **37**, 431–4.

Rasmus, K. (2009). A thermodynamics modelling study of an idealised low-elevation blue ice area in Antarctica. *Journal of Glaciology* **55**(194), 1083–91.

Rau, F. and Braun, M. (2002). The regional distribution of the dry snow zone on the Antarctic Peninsula north of 70°S. *Annals of Glaciology* **34**, 95–100.

Raymond, J., Zhaxybayeva, O., Gogarten, J.P., Gerdes, S.Y. and Blankenship, R.E. (2002). Whole-genome analysis of photosynthetic prokaryotes. *Science* **298**, 1616–20.

Reijmer, C.H. and Van den Broeke, M.R. (2003). Temporal and spatial variability of the surface mass balance in Dronning Maud Land, Antarctica, as derived from automatic weather stations. *Journal of Glaciology* **49**, 295–302.

Remais, D., Lütz-Meindl, U. and Lütz, C. (2005). Photosynthesis, pigments and ultrastructure of the alpine snow alga *Chlamydomonas nivalis*. *European Journal of Phycology* **40**, 259–68.

Reynolds, B. (1983). The chemical composition of snow at a rural upland site in mid-Wales. *Atmospheric Environment* **17**, 1849–51.

Reynolds, J. (2000). On the formation of supraglacial lakes on debris-covered glaciers. In Nakawo, M., Raymond, C.F. and Fountain, A. (eds) *Symposium at Seattle 2000: Debris-covered glaciers*, pp 153–61. International Association of Hydrological Sciences Publication No. 264, IAHS Press Oxfordshire, UK.

Richards, K., Sharp, M., Arnold, N., *et al.* (1996). An integrated approach to modelling hydrology and water quality in glacierized catchments. *Hydrological Processes* **10**, 479–508.

Ridley, J. (1993). Surface melting on Antarctic peninsula ice shelves detected by passive microwave sensors. *Geophysical Research Letters* **20**(23), 2639–42.

Riedel, A., Michel, C., Gosselin, M. and LeBlanc, B. (2007). Enrichment of nutrients, exopolymeric substances and microorganisms in newly formed sea ice on the Mackenzie shelf. *Marine Ecology Progress Series* **342**, 55–67.

Rieder, R., Economou, T., Wänke, H., *et al.* (1997). The chemical composition of martian soil and rocks returned by the mobile Alpha Proton X-ray Spectrometer: preliminary results from the X-ray mode. *Science* **278**, 1771–8.

Rignot, E., Bamber, J.L., Van den Broeke, M.R., *et al.* (2008). Recent Antarctic mass loss from radar interferometry and regional climate modelling, *Nature Geoscience* **1**, 106–10.

Riley, J., Hoppa, V., Greenberg, R., Tuffs, B.R. and Giessler, P. (2000). Distribution of chaotic terrain on Europa. *Journal of Geophysical Research* **105**, 22599–615.

Roberts, E.C. and Laybourn-Parry, J. (1999). Mixotrophic cryptophytes and their predators in the Dry Valley lakes of Antarctica. *Freshwater Biology* **41**, 737–46.

Roberts, T.J., Hodson, A.J., Evans, C.D. and Holmém, K. (2010). Modelling the impacts of a nitrogen pollution event on the biogeochemistry of an Arctic glacier. *Annals of Glaciology* **56**, 163–70.

Robin, G. de Q., Swithinbank, C.W.M. and Smith, B.M.E. (1970). Radio echo exploration of the Antarctica ice sheet. In Gow, A.J., Keeler, C., Langway, C.C. and Weeks, W.F. (eds) *International Symposium on Antarctic Glaciological Exploration, Hanover, New Hampshire, 3–7 September, 1968.* International Association of Scientific Hydrology Publication No. 86, IAHS Press, Oxfordshire UK.

Rochet, M., Legendre, L. and Demers, S. (1985). Acclimation of sea-ice microalgae to freezing temperature. *Marine Ecology Progress Series* **24**, 187–91.

Rózanska, M., Gosselin, M., Poulin, M., Wiktor, J.M. and Michel, C. (2009). Influence of environmental factors on the development of bottom ice protest communities during winter–spring transition. *Marine Ecology Progress Series* **386**, 43–59.

Ruiz, J., Montoya, L., López, V. and Amils, R. (2007). Thermal diaprisim and habitability of the icy shell of Europa. *Origin of Life, Evolution and the Biosphere* **37**, 287–95.

Runnegar, B. (2008). Loophole for snowball Earth. *Nature* **405**, 403–4.

Rutter, N., Essery, R., Pomeroy, J., *et al.* (2009). Evaluation of forest snow processes models (SnowMIP2). *Journal of Geophysical Research* **114**(D06111), doi 10.1029/2008JD011063.

Salonen, K., Leppäranta, M., Viljanen, M. and Gulati, R.D. (2009). Perspectives in winter limnology: closing the annual cycle of freezing lakes. *Aquatic Ecology* **43**, 609–16.

Sattler, B., Puxbaum, H. and Psenner, R. (2001). Bacterial growth in supercooled cloud droplets. *Geophysical Research Letters* **28**, 243–6.

Sattler, B., Storrie-Lombardi, M.C., Foreman, C., Tilg, M. and Psenner, R. (2010). Laser-induced fluorescence emission (LIFE) from Lake Fryxell (Antarctica) cryoconites. *Annals of Glaciology* **56**, 145–52.

Saunders, N.F.W., Thomas, T., Curmi, P.M.G., *et al.* (2003). Mechanisms of thermal adaptation revealed from the genomes of the Antarctic Archaea *Methanogenium frigidum* and *Methanococcoides burtonii. Genome Research* **13**, 1580–8.

Säwström, C., Mumford, P., Marshall, W., Hodson, A.J. and Laybourn-Parry, J. (2002). The microbial communities and primary productivity of cryoconite holes in an Arctic glacier (Svalbard 79°N). *Polar Biology* **25**, 591–6.

Säwström, C., Anesio, A.M., Granéli, W. and Laybourn-Parry, J. (2007a). Seasonal viral loop dynamics in two ultra-oligotrophic Antarctic freshwater lakes. *Microbial Ecology* **53**, 1–11.

Säwström, C., Granéli, W., Laybourn-Parry, J. and Anesio, A.M. (2007b). High viral infection rates in Antarctic and Arctic bacterioplankton. *Environmental Microbiology* **9**, 250–5.

Säwström, C., Laybourn-Parry, J., Anesio, A.M., Priscu, J.C. and Lisle, J. (2008). Viruses in polar inland waters. *Extremophiles* **12**, 167–75.

Scambos, T., Hulbe, C. and Fahnestock, M. (2003). Climate-induced ice shelf disintegration in the Antarctic Peninsula. In Domack, E., Burnett, A., Leventer, A., Conley, P., Kirby, M. and Bindschadler, R. (eds) *Antarctic Peninsula Climate Variability* pp 79–92. Antarctic Research Series Vol. 79. American Geophysical Union, Washingon, DC.

Schlosser, E., Powers, J.G., Duda, M.G. and Manning, K.W. (2011). Interaction between Antarctic sea ice and synoptic activity in the circumpolar trough: implications for ice-core interpretation. *Annals of Glaciology* **52**(57), 9–17.

Schmidt, S.K., Wilson, K.L., Monson, R.K. and Lipson, D.A. (2009). Exponential growth of 'snow molds' at subzero temperatures: an explanation for high beneath snow respiration rates and Q_{10} values. *Biogeochemistry* **95**, 12–21.

Schnack-Schiel, S.B., Dieckmann, G.S., Gradinger, R., Melnikov, I.A., Spindle, M. and Thomas, D.N. (2000). Meiofauna in sea ice of the Weddell Sea (Antarctica). *Polar Biology* **24**, 724–8.

Schrag, D.P. and Hoffman, P.F. (2001). Life, geology and snowball Earth. *Nature* **409**, 306.

Schultze-Makuch, D. and Irwin, L.N. (2002). Energy cycling and hypothetical organisms in Europa's ocean. *Astrobiology* **2**, 105–21.

Schünemann, H. and Werner, I. (2005). Seasonal variations in distribution patterns of sympagic meiofauna in Arctic pack ice. *Marine Biology* **146**, 1091–102.

Schwarz, J.N. and Schodlok, M.P. (2009). Impact of drifting icebergs on surface phytoplankton biomass in the Southern Ocean: ocean colour remote sensing and in situ iceberg tracking. *Deep Sea Research* **56**, 1727–41.

Scott, D., Hood, E. and Nassry, M. (2011). In-stream uotake and retention of C, N and P in a supraglacial stream. *Annals of Glaciology* **56**, 80–6.

Scott, F.J. and Marchant, H.J. (2005). *Antarctic Marine Protists*. ABRS, Canberra, and AAD, Hobart.

Segawa, T. and Takeuchi, N. (2010). A cyanobacterial community on the Qiyi Glacier in the Qilian Mountains of China. *Annals of Glaciology* **51**, 135–44.

Segawa, T., Miyamoto, K., Ushida, K., Agata, K., Okada, N. and Kohshima, S. (2005). Seasonal change in bacterial flora and biomass in mountain snow from the Tateyama Mountain, Japan, analysed by 16S rRNA gene sequencing ands real-time PCR. *Applied and Environmental Microbiology* **71**, 123–30.

Segura, A. and Navarro-Gonzalez, R. (2005). Nitrogen fixation on early Mars by volcanic lightening and other sources. *Geophyscial Research Letters* **32**, L05203.

Semkin, R.G. and Jeffries, D.S. (1988). Chemistry of atmospheric deposition, the snowpack and snowmelt in the Turkey Lakes Watershed. *Canadian Journal of Fisheries and Aquatic Sciences* **45**, 38–46.

Serreze, M.C. (2011). Climate change: rethinking the sea-ice tipping point. *Nature* **471**(7336), 47–810, doi 1038/471047a.

Serreze, M.C., Maslanik, J.A., Scambos, T.A., *et al.* (2003). A record minimum arctic sea ice extent and area in 2002. *Geophysical Research Letters* **30**(3), 1110.

Shain, D.H., Carter, M.R., Murray, K.P., *et al.* (2000). Morphologic characterization of the ice worm *Mesenchtraeus solifugus*. *Journal of Morphology* **246**, 192–7.

Sharp, M.J., Richards, K. Willis. I. *et al.* (1993). Geometry, bed topography and drainage system of the Haut Glacier d'Arolla, Switzerland. *Earth Surface Processes and Landforms* **18**, 557–71.

Sharp, M., Tranter, M., Brown, G. and Skidmore, M. (1995). Rates of chemical weathering and CO_2 drawdown in a glacier-covered alpine catchment. *Geology* **23**, 62–4.

Sharp, M., Parkes, J., Cragg, B., Fairchild, I.J., Lamb, H. and Tranter, M. (1999). Widespread bacterial populations at glacier beds and their relationship to rock weathering and carbon cycling. *Geology* **27**, 107–10.

Shaw, P.M., Russell, L.M., Jefferson, A. and Quinn, P.K. (2010). Arctic organic aerosol measurements show particles from mixed combustion in spring haze and from frost flowers in winter. *Geophysical Research Letters* **37**(10), L10803.

Shean, D.E., Head, J.W. and Marchant, D.R. (2005). Origin and evolution of a cold-based topical mountain glacier on Mars: the Pavonis Mons fan-shaped deposit. *Journal of Geophysical Research* **110**, doi 10.1029/2004JE00260.

Sheridan, P.P., Miteva, V.I. and Brenchley, J.E. (2003). Phylogenetic analysis of anaerobic psychrophilic enrichment cultures obtained from a Greenland glacier ice core. *Applied and Environmental Microbiology* **69**, 2153–60.

Shi, T., Reeves, R.H., Gilichinsky, D.A. and Freidmann, E.I. (1997). Characterization of viable bacteria from Siberian permafrost by 16S rRNA sequencing. *Microbial Ecology* **33**, 169–79.

Shindell, D.T., and Schmidt, G.A. (2004). Southern Hemisphere climate response to ozone changes and greenhouse gas increases. *Geophysical Research Letters* **31**, L18209, doi 10.1029/2004GL020724.

Siegal, V. and Loeb, V. (1995). Recruitment of Antarctic krill *Euphausia superba* and possible causes for its variability. *Marine Ecology Progress Series* **123**, 45–56.

Siegert, M. (2001). *Ice Sheets and Quaternary Change*. Wiley, Chichester, UK, 231 pp.

Siegert, M.J. and Dowdeswell, J.A. (1996). Spatial variations in heat at the base of the Antarctic ice sheet from analysis of the thermal regime above sub-glacial lakes. *Journal of Glaciology* **42**, 501–9.

Siegert, M.J., Tranter, M., Ellis-Evans, J.C., Priscu, J.C. and Lyons, W.B. (2003). The hydrochemistry of Lake Vostok and the potential for lifein Antarctic sub-glacial lakes. *Hydrological Processes* **17**, 795–814.

Siegert, M.J., Carter, S., Tabacco, I., Popov, S. and Blankenship, D.D. (2005). A revised inventory of Antarctic subglacial lakes. *Antarctic Science* **17**, 453–60.

Siegert, M.J., Behar, A., Bentley, M., *et al.* (2006). Exploration of Ellesworth Subglacial Lake: a concept paper on the development, organization and execution of an experiment to explore, measure and sample the environment of a West Antarctic subglacial lake, the Ellesworth Consortium. *Reviews in Environmental Science and Biotechnology* **6**, 161–79.

Siegert, M.J., Le Brocq, A. and Payne, A.J. (2007). Hydrological connections between Antarctic subglacial lakes, the flow of water beneath the East Antarctic ice sheet and implications for sedimentary processes. In Hambrey, H.J., Christoffersen, P., Glasser, N.F. and Hubbard, B. (eds) *Glacial Sedimentary Processes and Products*, pp 3–10. Wiley-Blackwell, Hoboken, NJ.

Sinensky, M. (1974). Homeoviscous adaptation—a homeostatic process that regulates the viscosity of the membrane lipids in *Esherichia coli*. *Proceeding of the National Academy of Sciences USA* **71**, 522–5.

Skidmore, M.L. and Sharp, M.J. (1999). Drainage system behaviour of a high Arctic polythermal glacier. *Annals of Glaciology* **28**, 209–15.

Skidmore, M.L., Foght, J.M. and Sharp, M.J. (2000). Microbial life beneath a high Arctic Glacier. *Applied and Environmental Microbiology* **66**, 3214–20.

Skidmore, M.L., Anderson, S.P., Sharp, M., Foght, J. and Lanoil, B. (2005). Comparison of microbial community compositions of two subglacial environments reveals a possible role for microbes in chemical weathering processes. *Applied and Environmental Microbiology* **71**, 6986–97.

Skidmore, M., Tranter, M., Tulaczyk, S. and Lanoil, B.D. (2010). Hydrochemistry of ice stream beds—evaporitic or microbial effects? *Hydrological Processes* **24**, 517–23.

Smith, A. (1997). Basal conditions on Rutford Ice Stream, West Antarctica, from seismic observations. *Journal of Geophysical Research* **102**(B1), 543–52.

Smith, B.E., Fricker, H.A., Joughin, I.R. and Tilaczyk, S. (2009). An inventory of active subglacial lakes in Antarctica detected by ICESat (2003–2008). *Journal of Glaciology* **55**, 573–95.

Smith, P.H., Tamppari, L.K., Aridson, R.E., *et al.* (2009). H_2O at the Phoenix landing site. *Science* **325**, 58–61.

Smith, R.E.H. and Clement, P. (1990). Heterotrophic activity and bacterial productivity in assemblages of microbes from sea ice in the High Arctic. *Polar Biology* **10**, 351–7.

Smith, R.E.H., Clement, P. and Cota, G.F. (1989). Population dynamics of bacteria in Arctic sea ice. *Microbial Ecology* **17**, 63–76.

Smith, W.O. and Nelson, D.M. (1986). Importance of ice edge phytoplankton production in the Southern Ocean. *Bioscience* **36**, 251–7.

Sneed, W.A. and Hamilton, G.S. (2007). Evolution of melt pond volume on the surface of the Greenland Ice Sheet. *Geophysical Research Letters* **34**, L03501.

Søreide, J.E., Leu, E., Berge, J., Graeve, M. and Falk-Petersen, S. (2010). Timing of blooms, algal food quality and *Calanus glacialis* preproduction and growth in a changing Arctic. *Global Change Biology* **16**, 3154–63.

Souchez, R., Petit, J.R., Jouzel, J., de Angelis, M. and Tison, J.L. (2004). Reassessing Lake Vostok's behaviour from existing and new ice core data. *Earth and Planetary Science Letters* **217**, 163–70.

Spaulding, S.A., McKnight, D.M., Smith, R.L. and Dufford, R. (1994). Phytoplankton population dynamics in perennially ice-covered Lake Fryxell, Antarctica. *Journal of Plankton Research* **16**, 527–41.

Spencer, J.S., Pearl, J.C., Segura, M., *et al.* (2006). Cassini encounters Enceladus: background and the discovery of a south polar hot spot. *Science* **311**, 1401–5.

Spohn, T. and Scubert, G. (2003). Oceans in the icy Galilean satellites of Jupiter? *Icarus* **161**, 456–67.

Squyres, S.W., Reynolds, R.T., Cassen, P.M. and Peale, S.J. (1983). The evolution of Enceladus. *Icarus* **53**, 319–31.

Squyres, S.W., Clifford, S.M., Kuzmin, R.O., Zimbelman, J.R. and Costard, F.M. (1992). Ice in the martian regolith. In Kiefer, H.H., Jakosky, B.M., Snyder, C.W. and Matthews, M.S. (eds) *Mars*, pp 523–54. Arizona University Press, Tuson, AZ.

Squyres, S.W., Knoll, A.H., Arvidson, R.E., *et al.* (2006). Two years at Meridiani Planum: results from the Opportunity Rover. *Science* **313**(5792), 1403–7.

Stahl, D.A., Lane, D.J., Olsen, G.J. and Pace N.R. (1985). Characterization of a Yellowstone hot spring microbial community by 5S rRNA sequences. *Applied and Environmental Microbiology* **49**, 1379–84.

Stal, L.S. (1995). Physiological ecology of cyanobacteria in microbial mats and other communities. *New Phytologist* **131**, 1–32.

Stein, J.L., Marsh, T.L., Wu, K.Y., Shizuya, H. and DeLong, E.F. (1996). Characterization of uncultivated prokaryotes: isolation and analysis of a 40 kilobase-pair genome fragment from a planktonic marine archaeon. *Journal of Bacteriology* **178**, 591–9.

Steinbock, O. (1936). Cryoconite holes and their biological significance. *Zeitschrift für Gletscherkunde* **24**, 1–21.

Steward, G.F., Smith, D.C. and Azam, F. (1996). Abundance and production of bacteria and viruses in the Bering and and Chukchi Seas. *Marine Ecology Progress Series* **131**, 287–300.

Stibal, M., Sabacká, M. and Kastovská, K. (2006). Microbial communities on glacier surfaces in Svalbard: impact of physical and chemical properties on abundance and structure of cyanobacteria and algae. *Microbial Ecology* **52**, 644–54.

Stibal, M., Tranter, M., Benning, L.G. and Rehák, J. (2008). Microbial primary production on an Arctic glacier is insignificant in comparison with allochthonous organic carbon input. *Environmental Microbiology* **10**, 2172–8.

Stibal, M., Lawson, E.C., Lis, G., Mak, K.M., Wadham, J.L. and Anesio, A.M. (2010). Organic matter content and quality in supraglacial debris across the ablation zone of the Greenland Ice Sheet, *Annals of Glaciology* **51**(56), 1–8.

Stoecker, D.K., Buck, K.R. and Putt, M. (1993). Changes in the sea–ice brine community during the spring summer transition, McMurdo Sound, Antarctica 11. Phagotrophic protists. *Marine Ecology Progress Series* **95**, 103–13.

Stoecker, D.K., Gustafson, D.E., Baier, C.T. and Black, M.M.D. (2000). Primary production in the upper sea ice. *Aquatic Microbial Ecology* **21**, 275–87.

Stohl, A., Berg, T., Burkhart, J.F., *et al.* (2007). Arctic smoke—record high air pollution levels in the European Arctic due to agricultural fires in Eastern Europe in spring 2006. *Atmospheric Chemistry and Physics* **7**, 511–34.

Storrie-Lombardi, M.C. and Sattler, B. (2009). Laser-induced fluorescence emission (LIFE): *in situ* nondestructive detection of microbial life in the ice covers of Antarctic lakes. *Astrobiology* **9**, 659–72.

Storrie-Lombardi, M.C., Muller, J-P., Fisk, M.R., Griffiths, A.D., Coates, A.J. and Hoover, R.B. (2008). Epifluorescence surveys of extreme environments using PanCam imaging systems: Antarctica and the Mars regolith. In Hoover, R.B., Levin, G.V., Rozanov, A.Y. and Davies, P.C. (eds) *Instruments, Methods, and Missions for Astrobiology XI*, pp 1–10. Society of Photo-Optical Instrumentation Engineers, Bellingham, WA.

Storrie-Lombardi, M.C., Muller, J-P., Fisk, M.R., *et al.* (2009). Laser induced fluorescence emission (LIFE): searching for Mars organics with a UV-enhanced PanCam. *Astrobiology* **9**, 953–64.

Stroeve, J., Holland, M.M., Meier, W., Scambos, T. and Serreze, M. (2007). Arctic sea ice decline: faster than forecast. *Geophysical Research Letters* **34**(9), L09501.

Stroeve, J.C., Maslanik, J., Serreze, M.C., Rigor, I., Meier, W. and Fowler, C. (2011). Sea ice response to an extreme negative phase of the Arctic Oscillation during winter 2009/2010. *Geophysical Research Letters* **38**(2), L02502.

Sugden, D.E. and John, B.S. (1984). *Glaciers and Landscape*. Arnold, London.

Summers, D.P. and Khare, B. (2007). Nitrogen fixation on early Mars and other terrestrial planets: experimental demonstration of abiotic fixation reactions to nitrite and nitrate. *Astrobiology* **7**, 333–41.

Sundal, A.V., Shephard, A., Nienow, P., *et al.* (2009). Evolution of supra-glacial lakes across the Greenland Ice Sheet. *Remote Sensing of Environment* **113**, 2164–71.

Sundal, A.V., Shepard, A., Nienow, P., Hanna, E., Palmer, S. and Huybrechts, P. (2011). Melt-induced speed-up of Greenland ice sheet offset by efficient drainage. *Nature* **469**, 521–4.

Suzuki, K. (1987). Spatial-distribution of chloride and sulfate in the snow cover in Sapporo, Japan. *Atmospheric Environment* **21**, 1773–8.

Szostak, J.W. (2009). Systems chemistry on early Earth. *Nature* **459**, 171–2.

Takacs, C.D. and Priscu, J.C. (1998). Bacterioplankton dynamics in the McMurdo Dry Valley lakes, Antarctica; production and biomass loss over four seasons. *Microbial Ecology* **36**, 239–50.

Takahashi, S., Endoh, T., Azuma, N. and Meshida, S. (1992). Bare ice fields developed in inland part of Antarctica. *Proceedings of the NIPR Symposium on Polar and Meteorological Glaciology* **5**, 128–39.

Takahashi, T., Tajiri, T. and Sonoi, Y. (1999). Charges on graupel and snow crystals and the electrical structure of winter thunderstorms. *Journal of Atmospheric Sciences* **56**, 1561–78.

Takeuchi, N. (2001). The altitudinal distribution of snow algae on an Alaska glacier (Gulkana Glacier in the Alaska Range). *Hydrological Processes* **15**, 3447–59.

Takeuchi, N. (2002). Optical characteristics of cryoconite (surface dust) on glaciers: the relationship between light absorbancy and the property of organic matter contained within the cryoconite. *Annals of Glaciology* **34**, 409–14.

Takeuchi, N., and Li, Z.Q. (2010). Characteristics of surface dust on Urumqi Glacier No. 1 in the Tien Shan Mountains, China. *Arctic Antarctic and Alpine Research* **40**(4), 744–50.

Takeuchi, N., Kohshima, S. and Fujita, K. (1998). Snow algae community on a Himalayan glacier, Glacier AX010, East Nepal: relationship with glacier summer mass balance. *Bulletin of Glacier Research* **16**, 43–50.

Takeuchi, N., Kohshima, S., Yoshimura, Y., Seko, K. and Fujita K. (2000). Characteristics of cryoconite holes on a Himalayan glacier, Yala Glacier, Central Nepal. *Bulletin of Glacier Research* **17**, 51–9.

Takeuchi, N., Kohshima, S. and Seko, K. (2001). Structure, formation, and darkening process of albedo-reducing material (cryoconite) on a Himalayan glacier: a granular algal mat growing on the glacier. *Arctic Antarctic and Alpine Research* **33**(2), 115–22.

Takeuchi, N., Dial, R., Kohshima, S. Segawa, T. and Uetake, J. (2006). Spatial distribution and abundance of red snow algae on the Harding Icefield, Alaska derived from a satellite image. *Geophysical Research Letters* **33**, L21502.

Takeuchi, N., Nishiyama, H. and Li, Z. (2010). Structure and formation process of cryoconite granules on Ürümqi glacier No. 1, Tien Shan, China. *Annals of Glaciology* **51**(56), 9–14.

Tang, E.P.Y., Tremblay, R. and Vincent, W.F. (1997). Cynaobacterial dominance of polar freshwater ecosystems: are high-latitude mat-formers adapted to low temperature? *Journal of Phycology* **33**, 171–81.

Taton, A., Grubisic, S., Bathazart, P., Hodgson D.A., Laybourn-Parry, J. and Wilmotte, A. (2006). Biogeographic distribution and ecological range of benthic cyanobacteria in East Antarctic lakes. *FEMS Microbiology Ecology* **57**, 272–89.

Taylor, G. (1922). *The Physiography of the McMurdo Sound and Granite Harbour Region*. Harrison and Sons, London.

Taylor Perron, J., Mitrovica, J.X., Manga, M., Matsuyama, I. and Richards, M.A. (2007). Evidence for an ancient martian ocean in the topography of deformed shorelines. *Nature* **447**, 840–3.

Tedesco, L., Vichi, M., Haapala, J. and Stipa, T. (2010). A dynamic biologically active layer for numerical studies of the sea ice ecosystem. *Ocean Modelling* 35, 89–104.

Telling, J., Anesio, A.M., Hawkings, J., *et al.* (2010). Measuring rates of gross photosynthesis and net community production in cryoconite holes: a comparison of field methods. *Annals of Glaciology* 51(56), 153–62.

Thomas, D.N. and Dieckmann, G.S. (2002). Antarctic sea ice—a habitat for extremophiles. *Science* 295, 641–4.

Thomas, D.N. and Papadimitriou, S. (2003). Biogeochemistry of sea ice. In Thomas, D.N. and Dieckmann, G.S. (eds) *Sea Ice—An introduction to its physics, chemistry, biology and geology*, pp 267–302. Blackwell, Oxford, UK.

Thomas, D.N., Lara, R.J., Eicken, H., Kattner, G. and Skoog, A. (1995). Dissolved organic matter in Arctic multi-year sea ice during winter: major components and relationships to ice characteristics. *Polar Biology* 15, 477–83.

Thomas, D.N., Lara, R.J., Hass, C., *et al.* (1998). Biological soup within decaying summer sea ice in the Amundsen Sea, Antarctica. In Lizotte, M.P. and Arrigo, K.R. (eds) *Antarctic Sea Ice Biological Processes, Interactions and Variability*, pp 161–71. Antarctic Research Series No. 73. American Geophysical Union, Washington, DC.

Thurman, J., Parry, J.D., Hill, P.J. and Laybourn-Parry, J. (2010). The filter-feeding ciliates *Colpidium* sp. and *Tetrahymena pyriformis* display selective feeding behaviours in the presence of mixed, equally-sized bacterial prey. *Protist* 161, 577–88.

Tietsche, S., Notz, D., Jungclaus, J.H. and Marotzke, J. (2011). Recovery mechanisms of Arctic summer sea ice. *Geophysical Research Letters* 38(2), L02707.

Titus, T.N., Kieffer, H.H. and Christensen, P.R. (2003). Exposed water ice discovered near the South Pole of Mars. *Science* 299, 1048–51.

Tokano, T., McKay, C.P., Neubauer, F.M., *et al.* (2006). Methane drizzle on Titan. *Nature* 442, 432–5.

Torinesi, O., Fily, M. and Genthon, C. (2003). Variability and trends of the summer melt period of Antarctic ice margins since 1980 from microwave sensors. *Journal of Climate* 16(7), 1047–60.

Tranter, M. (2004). Geophysical weathering in glacial and proglacial environments. In Drever, J.I. (ed.) *Treatise on Geochemistry, Vol. 5. Surface and ground water, weathering and soils*, pp 189–206. Elsevier, Oxford, UK.

Tranter, M., Brimblecombe, P., Davies, T.D., Vincent, C.E., Abrahams, P.W. and Blackwood, I. (1986). The composition of snowfall, snowpack and meltwater in the Scottish Highlands—evidence for preferential elution. *Atmospheric Environment* 20, 517–25.

Tranter, M., Abrahams, P.W., Blackwood, I.L., Brimblecombe, P. and Davies, T.D. (1988). The imact of a

single black snowfall on streamwater chemistry in the Scottish Highland. *Nature* 332, 826–9.

Tranter, M., Sharp, M.J., Lamb, H.R., Brown, G.H., Hubbard, B.P. and Williss, I.C. (2002a). Geochemical weathering at the bed of Haut Glacier d'Arolla, Switzerland—a new model. *Hydrological Processes* 16, 959–93.

Tranter, M., Huybrechts, P., Sharp, M.J., *et al.* (2002b). Direct effect of ice sheets on terrestrial bicarbonate, sulphate, and base cation fluxes during the last glacial cycle: minimal impact on atmospheric CO_2 concentrations. *Chemical Geology* 190, 33–44.

Tranter, M., Fountain, A., Fritsen, C., *et al.* (2004). Extreme hydrochemical conditions in natural microcosms entombed within Antarctic ice. *Hydrological Processes* 18, 379–87.

Tranter, M., Skidmore, M.L. and Wadham, J. (2005). Hydrological controls on microbial communities in subglacial environments. *Hydrological Processes* 19, 996–8.

Tranter, M., Bagshaw, E.M., Fountain, A.G. and Foreman, C.M. (2010). The biogeochemistry and hydrology of McMurdo Dry Valley glaciers: is there life on martian ice now? In Doran, P.T., Lyons, W.B. and McKnight, D.M. (eds) *Life in Antarctic Deserts and Other Cold, Dry Environments: Astrobiological analogues*, pp 195–220. Cambridge University Press, Cambridge, UK.

Tulaczyk, S. and Hossainzadeh, S. (2011). Antarctica's deep frozen 'lakes'. *Science* 331, 1524–5.

Tulaczyk, S., Kamb, W.B. and Engelhardt, H.F. (2000). Basal mechanics of Ice Stream B, West Antarctica 2. Undrained plastic bed model. *Journal of Geophysical Research* 105, 483–94.

Tung, H.C., Brammell, N.E. and Price, P.B. (2005). Microbial origin of excess methane in glacial ice and implications for life on Mars. *Proceedings of the National Academy of Sciences* 102, 18292–6.

Tung, H.C., Price, P.B., Bramell, N.E. and Vrdoljak, G. (2006). Microorganisms metabolizing on clay grains in 3-km-deep Greenland basal ice. *Astrobiology* 6, 69–86.

Turner, J., Comiso, J.C., Marshall, G.J., *et al.* (2009). Non-annular atmospheric circulation change induced by stratospheric ozone depletion and its role in the recent increase of Antarctic sea ice extent. *Geophysical Research Letters* 36(8), L08502.

Turtle, E.P. and Pierazzo, E. (2001). Thickness of a Europan ice shell from impact crater simulations. *Science* 294, 1326–8.

Tye, A.M., Young, S.D., Crout, N.M.J., *et al.* (2005). The fate of ^{15}N added to high Arctic tundra to mimic increased inputs of atmospheric nitrogen released from a melting snowpack. *Global Change Biology* 11, 1640–54.

Underwood, G.J.C., Fietz, S., Papadimitriou, S., Thomas, D.N. and Dieckmann, G.S. (2010). Distribution and composition of dissolved extracellular polymeric substances (EPS) in Antarctic sea ice. *Marine Ecology Progress Series* **404**, 1–19.

Vaughan, D.G. (2006). Recent trends in melting conditions on the Antarctic Peninsula and their implications for ice-sheet mass balance. *Arctic, Antarctic and Alpine Research* **38**, 147–52.

Vaughan, D.G., Marshall, G.J., Connolley, W.M., *et al.* (2003). Recent rapid regional climate warming on the Antarctic Peninsula. *Climate Change* **60**(3), 243–74.

Villar, S.E.J. and Edwards, H.G.M. (2006). Raman spectroscopy in astrobiology. *Analytical and Bioanalytical Chemistry* **384**, 100–13.

Vincent, W.F. (1988). *Microbial Ecosystems in Antarctica.* Cambridge University Press, Cambridge, UK, 304 pp.

Vincent, W.F. and Howard-Williams, C. (2000). Life on Snowball Earth. *Science* **287**, 2421.

Vincent, W.F., Catenholz, R.W., Downes, M.T. and Howard-Williams, C. (1993). Antarctic cyanobacteria: light, nutrients and photosynthesis in the microbial mat environment. *Journal of Phycology* **29**, 745–55.

Vincent, W.F., Gibson, J.A., Pienitz, R., *et al.* (2000). Ice shelf microbial ecosystems in the high Arctic and implications for life on snowball earth. *Naturwissenchaften* **87**, 137–41.

Vincent, W.F., Mueller, D.R. and Bonilla, S. (2004). Ecosystems on ice: the microbial ecology of Markham Ice Shelf in the high Arctic. *Cryobiology* **48**, 103–12.

Virginia, R.A. and Wall, D.H. (1999). Soil animals in an extreme environment: how soils structure communities in the McMurdo Dry valleys, Antarctica. *Bioscience* **49**, 973–83.

Wadham, G., Botterell, S., Tranter, M. and Raiswell, R. (2004). Stable isotope evidence for microbial sulphate reduction at the bed of a polythermal high Arctic glacier. *Earth and Planetary Science Letters* **219**, 341–55.

Wadham, J.L. and Nuttall, A-M. (2002). Multiphase formation of superimposed ice during a mass-balance year at a maritime high-Arctic glacier. *Journal of Glaciology* **48**(163), 545–51.

Wadham, J.L., Hallam, K.R., Hawkins, J.M. and O'Connor, A. (2006). Enhancement of snowpack inorganic nitrogen by aerosol debris. *Tellus, B: Chemical and Physical Meteorology* **58**(3), 229–41.

Wadham, J.L., Tranter, M., Tulaczyk, S. and Sharp, M. (2008). Subglacial methanogenesis: a potential climate amplifier? *Global Biogeochemical Cycles* **22**, doi 10.1029/2007gb002951.

Wadham, J.L., Tranter, M., Skidmore, M., *et al.* (2010). Biogeochemcial weathering under ice: size matters. *Global Biogeochemcial Cycles* **24**(3), GB3025.

Wager, A.C. (1972). Flooding of the ice shelf in George VI Sound. *British Antarctic Survey Bulletin* **28**, 71–4.

Wagnon, P., Ribstein, P., Kaser, G. and Berton, P. (1999). Energy balance and runoff seasonality of a Bolivian glacier. *Global and Planetary Change* **22**(1–4), 49–58.

Wait, B.R., Nokes, R. and Webster-Brown, J.G. (2009). Freeze–thaw dynamics and the implications for stratification and brine geochemistry in meltwater ponds on the McMurdo Ice Shelf, Antarctica. *Antarctic Science* **21**, 243–54.

Waleed, A. and Steffen, K. (1997). Snowmelt on the Greenland Ice Sheet as derived from passive microwave satellite data. *Journal of Climate* **10**, 165–75.

Wang, M. and Overland, J.E. (2009). A sea ice free summer Arctic within 30 years? *Geophysical Research Letters* **36**(7), L07502.

Warren, S.G. and Hudson, S.R. (2003). Letters to the editor—Bacterial activity in South Pole snow is questionable. *Applied and Environmental Microbiology* **69**, 6340–1.

Wells, M.G. and Wettlaufer, J.S. (2008). Circulation in Lake Vostok: a laboratory analogues study. *Geophysical Research Letters* **35**, doi 10.1029/2007GL032162.

Weinbauer, M.G. (2004). Ecology of prokaryote viruses. *FEMS Microbial Reviews* **28**, 127–81.

Weinbauer, M.G. and Suttle, C.A. (1996). Potential significance of lysogeny to bacteriophage production and bacterial mortality in coastal waters of the Gulf of Mexico. *Applied and Environmental Microbiology* **62**, 4374–80.

Werner, I. and Gradinger, R. (2002). Under-ice amphipods in the Greenland Sea and Fram Strait (Arctic): environmental controls and seasonal patterns below pack ice. *Marine Biology* **140**, 317–26.

Werner, I., Ikävalko, J. and Schünemann, H. (2007). Sea-ice algae in Arctic pack ice during late winter. *Polar Biology* **30**, 1493–504.

Wharton, R.A., Vinyard, W.C., Parker, B.C., Simmins, G.M. and Seaburg, K.G. (1981). Algae in cryoconite holes on the Canada Glacier in southern Victoria Land. *Phycologia* **20**, 208–11.

Whittaker, R.H. (1969). New concepts of kingdoms of organisms. *Science* **163**, 150–60.

Wiacek, A. and Peter, T. (2009). On the availability of uncoated mineral dust ice nuclei in cold cloud regions. *Geophysical Research Letters* **36**, L17801, doi 17810.11029/12009 GL039429.

Wilch, E. and Hughes, T.J. (2000). Calculating basal thermal zones beneath the Antarctic ice sheet. *Journal of Glaciology* **46**(153), 297–310.

Wilhelm, S.W., Jeffery W.H., Dean A.L., Meador J., Pakulski J.D. and Mitchell D.L. (2003). UV radiation induced DNA damage in marine viruses along a latitudinal gradient in the southeastern Pacific Ocean. *Aquatic Microbial Ecology* **31**, 1–8.

Wilkins, M.R., Appel, R.D., Van Eyk, J., *et al.* (2006). Guidelines for the next 10 years of proteomics. *Proteomics* **6**, 4–8.

Willerslev, E., Cappellini, E., Boomsma, W., *et al.* (2007). Ancient biomolecules from deep ice cores reveal a forested southern Greenland. *Science* **317**, 111–14.

Williams, M.W., Seibald, C. and Chowanski, K. (2009). Storage and release of solutes from a subalpine seasonal snowpack: soil and stream water responses, Niwot Ridge, Colorado. *Biogeochemistry* **95**, 77–94.

Wingham, D.J., Siegert, M.J., Shepherd, A. and Muir, A.S. (2006). Rapid discharge connects Antarctic subglacial lakes. *Nature* **440**, 1033–6.

Winther, J.G., Jespersen, M.N. and Liston, G.E. (2001). Blue ice areas in Antarctica derived from NOAA AVHRR satellite data. *Journal of Glaciology* **47**(157), 325–34.

WMO. (2009). *Sea Ice Nomenclature.* World Meterological Organization, Geneva, Switzerland.

Woese, C.R. (1987). Bacterial evolution. *Microbiology Reviews* **51**, 221–71.

Wolff, E.W. and Paren, J.G. (1984). A two-phase model of electrical conduction in polar ice sheets. *Journal of Geophysical Research* **89**, 9433–8.

Wolken, G.J., Sharp, M. and Wang, L. (2009). Snow and ice facies variability and ice layer formation on Canadian Arctic ice caps, 1999–2005. *Journal of Geophysical Research* **114**(F03011), doi 10.1029/2008JF001173.

Wommack, K.E. and Colwell, R.R. (2000). Viriopkankton: viruses in aquatic ecosystems. *Microbiological and Molecular Biology Reviews* **64**, 69–114.

Wommack, K.E., Hill, R.T., Muller, T.A. and Colwell, R.R. (1996). Effects of sunlight on bacteriophage viability and structure. *Applied and Environmental Microbiology* **62**, 1336–41.

Woodin, S.J. (1997). Effects of acid deposition on Arctic vegetation. In Woodin, S.J. and Marquiss, M (eds) *Ecology of Arctic Environments*, pp 190–239. Blackwell Publications, Oxford, UK.

Wright, A. and Siegert, M.J. (2011). The identification and physiographical setting of Antarctic subglacial lakes: an update based on recent geophysical data. In Siegert, M., Kennicutt, C. and Bindschadler, B. (eds) *Subglacial Antarctic Aquatic Environments.* American Geophysical Union Monograph. American Geophysical Union, Washington, DC, pp 9–26.

Wynn, P.M., Hodson, A. and Heaton, T. (2006). Chemical and isotopic switching within the subglacial environment of a High Arctic glacier. *Biogeochemistry* **78**, 173–93.

Yackel, J.J., Barber, D.G., Papakyriakou, T.N. and Breneman, C. (2007). First-year sea ice spring melt transitions in the Canadian Arctic Archipelago from time-series synthetic aperture radar data, 1992–2002. *Hydrological Processes* **21**(2), 253–65.

Yallop, M.L. and Anesio, A.M. (2010). Benthic diatom flora in supra-glacial habitats: a generic level comparison. *Annals of Glaciology* **51**(56), 15–22.

Yang, X.X., Lin, X.Z., Bian, J., Sun, X.Q. and Hunag, X.H. (2004). Identification of five strains of Antarctic bacteria producing low temperature lipase. *Acta Oceanology Sinica* **23**, 717–23.

Yde, J.C., Riger-Kusk, M., Christiansen, H.H., Knudsen, N.T. and Humlum, O. (2008). Hydrochemical characteristics of bulk meltwater from an entire alblation season, Longyearbyen, Svalbard. *Journal of Glaciology* **54**, 259–72.

Yde, J.C., Finster, R.W., Raiswell, R., *et al.* (2010). Basal ice microbiology at the margin of the Greenland Ice Sheet. *Annals of Glaciology* **51**, 71–9.

Yoshimura, Y., Kohshima, S. and Ohtani, S. (1997). A community of snow algae on a Himalayan glacier: change of algal biomass and community structure with altitude. *Arctic and Alpine Research* **29**, 126–37.

Yoshimura, Y., Kohshima, S., Takeuchi, N., Seko, K. and Fujita, K. (2000). Himalayan ice-core dating with snow algae. *Journal of Glaciology* **46**(153), 335–40.

Zhang, J., Steele, M. and Schweiger, A. (2010). Arctic sea ice response to atmospheric forcings with varying levels of anthropogenic warming and climate variability. *Geophysical Research Letters* **37**(20), L20505.

Zimmerman, W. and Feldman, J. (2002). Cryobot: and ice penetrating robotic vehicle for Mars and Europa. *Proceedings of the IEEE Aerospace Conference 10th March 2001–17th March 2001* **1**, 311–23.

Index

ablation 20, 47, 54, 140
accretion ice 113–115
accumulation 20
Accumulation Area Ratios (AAR) 20
Acidobacteria- *Holophaga* 19
Actinobacteria 19, 31, 40, 111, 115, 119
adaptation 77–78
Adenosine 5′ phosphosulphate reductase (APR) 109
Admiralty Bay 129
aerosols 31, 37
aggregates 59–60, 65, 73–74, 95
airborne platforms 133
Algae 2, 8, 13, 53
    see also diatoms; desmids; chlorophytes
algal blooms 134
algal mats 48–52
    see also cyanobacterial mats
alkaline phosphatase 41
allochthonous carbon 18, 104
Alphaproteobacteria 31, 62, 111, 115
altimetry 21
Amery Ice Shelf
    among largest 47
    exploration 133
amphipods 89–91
ammonium (NH$_4$-N) 39–40, 44–45, 53, 140
amoebae 7, 78, 87
Anabaena 50
anaerobic nitrogen reducers 30
Ancylonema nordenskiöldii 61
annual lake ice 18, 96–99
Antarctica 16, 19–20, 32
Antarctic Ice Sheet
    basal temperature 27
    basal water 101
    ice coring 130–131
    Lake Vostok 103
    melting snow surfaces 29
    subglacial lakes 102
Antarctic lakes 17–18
Antarctic Peninsula 20–21
Antarctic sea ice 74–75

anthropogenic activity 37
antifreeze protein 40
Apherusa glacialis 89–90
Arabia Terra (Mars) 125
Archaea 6–7, 104, 107, 110–111, 119, 141
Arctic Ocean 73–74, 92
Arthrobacter 111
Askenasia 18
Asplanchna 97
assimilation number 69, 82, 96
atmospheric deposition 37
atmosphere on Europa 125
atmosphere on Titan 127
Aulacoseira baicalensis 99
Aulacoseira skvortzowii 99
Austre Brøggerbreen
    ammonium assimilation 53
    bacteria in cryoconite holes 63
    bacterial production 65
    community respiration 70
    photosynthesis 66
autonomous platforms 130, 134
autotrophs 2–3

Bacteria 2–4, 6–7, 13, 15, 35, 61–63, 141
    chemoautotrophic bacteria 7, 30, 104, 115–116
    chemolithotrophic bacteria 128
    chemotrophic bacteria 109
    filamentous bacteria 83
    green filamentous bacteria 142
    green non-sulphur bacteria 18, 142
    heterotrophic bacteria 7, 52, 62–63, 104
    iron reducing bacteria 111
    Martian bacteria 122
    methanogenic bacteria 30
    methylotrophic bacteria 115
    photosynthetic bacteria 7, 142
    purple bacteria 142
    sulphate reducing bacteria 48
bacterial biomass 40, 53
bacterial concentrations 30, 35, 40,

54, 56–57, 62–63, 71, 85, 95, 97, 105, 109, 118–119, 132
bacterial growth (production) 30, 52, 57, 64–65, 83–85, 95, 98, 109, 118
bacterial growth efficiency (BGE) 70
bacterial morphology 16
bacterial phylotypes 109, 119
bacterial size 64–65
Bacteriodetes 31, 62, 115, 119
Beaver Lake
    chlorophyll *a* 61
    photosynthetic efficiency 68
    photosynthetic rate 66
Barents Ice Sheet
    Last Glacial Maximum 9
basal ice 20
basal melting 26
basal till 102
Bench Glacier
    bacterial concentrations 109
    bacterial diversity 106
Betaproteobacteria 30–31, 62, 98, 105–106, 109, 111, 115
Bindschadler ice steam 114
biofilms 59–60
biogeochemical cycling 5, 63
biogeography of cyanobacteria 51
bioinformatics 140
Blood Falls 107–109
blue ice areas (BIA) 22, 54, 133
Bracteococcus 49
brine 117, 119
brine channels 13, 77, 140
Burkholderia 42

calanoid copepods 89–92
Callisto 125
calving loss 20, 47
calyptopsis 90–91
Commonwealth Glacier
    bacterial concentrations 63
    rate of bacterial production 65
    cyanobacteria 56
    invertebrates 57

Canada Glacier
  bacterial concentrations 63
  rate of bacterial production 65
  chlorophyll *a* 61
  community respiration 70
  cyanobacteria 56
  invertebrates 57
  interlinked valleys 55
  photosynthetic rate 66
cap carbonates 10
capsular polysaccharides (cps) 49
carbon 4–5, 39
[13]carbon 10
[14]carbon 10
carbon balance models 139
carbon cycling 63–71
carbon dioxide 3, 12, 37, 41–42, 66,
    69, 103–104, 108, 139
carbon dioxide jets 124
carbon stable isotopes 41
carbonates 103
carbonation 103
carotenoids 49–50, 77
cold based glaciers 25–26
cold ice thermal regime 25
Chaetocerus 80
channelized drainage systems 102
Char Lake
  chlorophyll *a* 61
  photosynthetic rate 66
chemoheterotrophs 30
Chlamydomonas 48, 119
Chlamydomonas nivalis 18, 35, 38, 61, 98
Chlorokybus 49
Chloromonas 77
Chloromonas alpine 35
Chloromonas nivalis 35
Chloromonas pichinchae 39
Chlorosarcinopsis 49
chlorophyll *a* 15, 18, 49–51, 58,
    61–62, 75, 77, 79, 82–83, 95, 97,
    99, 118–119
chlorophyll *b* 49–50
chlorophyll mapping 133
chloroplasts 3
chlorophytes 96, 119 *see also Algae*
Chromulina 98
Chroococcales 61
Chroomonas 48
chrysophytes 3, 96–97, 119
    see also phytoflagellates
ciliates 3–4, 8, 54, 63, 78, 86–87, 97,
    99, 119
Ciliophora 8
clathrates 115, 137
climate driven surface mass balance
    model 21
climate change/warming 1, 9, 11,

13–14, 20–21, 42, 47, 55, 73–75,
    90, 92–94, 107, 130, 139
cold adaptation protein 141
cold-based areas 101
cold shock proteins 40
colony forming units (CFUs) 111
Colour Lake
  residual ice cover 94
columnar ice 13, 77
Colwellia 83–84
Colwellia psychreythraea 88
Comamonas 106
community structure 2–6
congelation growth 74
comparative genomics 141–142
copepodid 90–91
copepods 78, 83, 89–91
Crooked Lake
  photosynthetic rate 66
  viral induced bacterial mortality 6
Crustacea 89–92
Cryobacterium psychrophilum 36
Cryobot 134–135
cryolakes 55
cryoconite 23, 30, 52, 57–60, 96
cryoconite communities 60–67
cryoconite holes 23, 30, 53–55,
    58–70, 134
Cryosat II Mission 133
cryptophytes 3–4, 6, 81–82, 87,
    96–97 *see also* phytoflagellates
crystallization of water 32
Cyanobacteria 3, 6–8, 18–19, 45–52,
    56–57, 60–61, 71, 95–96, 99, 119,
    142
cyanobacterial mats 48–52
cyclopoid copepods 89–92
Cylindrocystis 61
Cylindrocystis brébissonii 39, 53
Cylindriotheca closterium 81
Cytophaga-Flavobacteria 30, 62, 84,
    98
Cytophaga-Flavobacterium-
    Bacteriodes 106, 109
Cytophaga-Flexibacteria-
    Bacteriodes 83–84
DAPI (4′, 6-diamidino-2-
    phenylindole) 2
$\delta^{13}C$ 10, 104
$\delta^{18}O$ 71, 104
$\delta^{18}O\text{-}SO_4^{2-}$ 108
$\delta^{18}O\text{-}H_2O$ 108
$\delta^{34}S\text{-}SO_4^{2-}$ 108

Deep Lake
  hypersalinity 120
  temperature 120
deglaciation 8–9

Deltaproteobacteria 106, 109
Denatured Gradient Gel
    Electrophoresis (DGGE) 119
denitrification 51
desmids 3, 45
Desulfocapsa sulflexigens 109
Diaphanoeca grandis 6
Diatoma 99
diatoms (Bacillariophyceae) 3, 8, 10,
    14–15, 45, 48, 49, 61, 78–81, 99,
    119
Didinium 15
Dileptus 18
dimethylsulfonioproprionate
    (DMSP) 78
dimictic lakes 94
Dinobryon 4
dinoflagellates 3, 16, 78, 81–82, 86
    see also phytoflagellates
Dileptus 97
distributed drainage system 101
dissolved organic carbon (DOC)
    4–6, 15–16, 35, 64, 74, 82–83, 95,
    104, 115, 118, 137–138
Drescheriella glacialis 89–91

East Antarctic Ice Sheet
  ablation 20
  accumulation 20
  community respiration 71
  Lake Vostok 112
  photosynthetic rate 71
  thickness 24
eH 101, 108
Elysium Planitia 123
Enceladus 126–127
energy budget 54
englacial 27, 29
Epsilonproteobacteria 106
euglenoids 3
    see also phytoflagellates
euphausids 89–91
Europa 125–126, 134
Europan oceans 126
exopolymer particles (EPs) 83
extracellular polymeric substances
    (EPS) 15, 49, 59, 83

facultative anaerobes 7
fast sea ice 73–76
Finsterwalderbreen
  bed anoxia 107
  sulphate reduction 104
Firmicutes 31, 40, 111
firnification 20
flagellated protozoans 2, 8, 78
    see also heterotrophic
      nanoflagellates; phytoflagellates

Flavobacterium hibernum 88
flow cytometry 137
flow fingers 33
flow rates 53
fluorescence *in situ* hybridization
  (FISH) 84
fluorescent dyes (fluorochromes) 2, 4
fluorescent microscopy 2, 4, 137
fluorescently labelled bacteria
  (FLBs) 86
Formaminifera 87–88, 91
Fox Glacier
  bacterial diversity 105–106
Fragilaria 99
Fragilariopsis curta 79
Fragilariopsis cylindrus 79–81
Fragilariopsis oceanica 79
Fram Strait 84
Franz Joseph Glacier
  bacterial diversity 105–106
frazil ice 73–74, 94
Frøya Glacier
  photosynthetic rate 66
fucoxanthin 15
Fungi 35, 42–44

Gammaproteobacteria 31, 40, 84,
  109, 111, 115, 119
Gammerus wilkitzkii 89–90
Ganymede 125
Geopsychrobacter electrodiphilus 109
Genomes 141–142
George VI Ice Shelf
  among largest 47
  exploration 133
geothermal heating 102–103, 111,
  117, 126
glacial biome 28
Glacier No1
  aggregates 60
glaciers 2, 8–12, 19–21, 23–25, 26–27,
  30–31, 47
  glaciers Alaskan 11
  glaciers alpine 9
  glacier beds 25–27
  glaciers Himalayan 53
  glacier mass balance 20
  glacier retreat 9–10
Glaciers
  Austre Brøggerbreen 53, 63,
    65–66, 70
  Bench Glacier 106, 109
  Canada Glacier 17, 55, 56–57, 61,
    63, 65–66, 70
  Commonwealth Glacier 56–57,
    63, 65
  Finsterwalderbreen 104, 107
  Fox Glacier 105–106

Franz Joseph Glacier 105–106
Frøya Glacier 66
Glacier No1 60
Glukana Glacier 54
Haut Glacier d'Arolla 25
Howard Glacier 56–57
Hughes Glacier 57, 63, 65
John Evans Glacier 106–107
Kangerlussuaq 66, 70
Longyearbreen 66, 70–71
Midtre Lovénbreen 53–54, 56–57,
  63, 65–66, 68, 70, 134
Qiyi Glacier 56
Rotmoosferner Glacier 63, 65
Stubacher Sonnblickkees 63,
  65–66, 70
Taylor Glacier 57, 107–108
Thwaites Glacier 129
Tuva Glacier 53, 71
Vestre Brøggerbreen 66, 70
Werenskiolbreen 64, 71
White Glacier 21, 61
Worthington Glacier 20
Glaciecola 84
Gloeocapsa 49
global glaciation 10–12
Glukana Glacier
  algal diversity 54
Gossenköllesee
  bacterial communities 18
  bacterial concentrations 98
  bacterial morphology 97
  bacterial production 98
  ice cover 18, 97–98
Greater Lake Priscu
  subglacial lake 107
Greenland Ice Sheet
  accumulation areas 32
  bacterial concentration 58
  bacterial diversity 111
  basal water 101
  carbon cycling 71
  carbon transfer 139
  chlorophyll *a* 58
  surface debris 71
  ice cores 130–132
  inorganic carbon 58
  melt zones 21–22, 134, 139
  organic carbon 58
  snow melt 29
  surface transition 52
  supraglacial lakes 23, 55–56, 102
Greenland Ice Sheet Project
  (GISP) 110–111, 130–132
ground penetrating radar (GPR) 27, 117
Guliya Ice Cap 111
Gymnodinium 86
Gyrodinium 86

Haloarchaea 7
Halomonas 84
Halteria 63
harpacticoid copepods 89–90
Haut Glacier d'Arolla
  moulins 25
Heliobacteria 142
heliozoans 78, 87
heterocysts 50–51
heterotrophic nanoflagellates
  grazing impact 86
heterotrophic nanoflagellates 3–4,
  15–16, 30, 36, 63, 78, 85–87, 97
heterotrophic nanoflagellates
  concentrations 63, 85
High Performance Liquid
  Chromatography (HPLC) 50–51
Holophrya 18
horizontal gene transfer (HGT)
  141–142
Howard Glacier
  cyanobacteria 56
  invertebrates 57
Hudson Bay, fast ice 82
Hughes Glacier
  bacterial cell size 65
  bacterial concentrations 63
  invertebrates 57
hydrogenation on Titan 127
Hydrogenophiulus thermoluteolus 115
hydrological transport 25
hydrological zonation 21–23
hydrothermal vents 126
hyporheic zone 28

ice active substances 40
ice algae 15–16
ice berg production 20
ice caps 102
ice edge blooms 76
ice granular 76
ice lids 60
ice margin 10
ice nuclei 31
ice nucleation 31
ice sheets
  Antarctic 27, 29, 101–103, 130, 131
  Barents 9
  East Antarctic 20, 24, 71, 112
  Greenland 21–23, 29, 32, 52,
    55–56, 58, 71, 101–102, 111,
    130–132, 134, 139
  Laurentide 9
  Patagonian 9
  West Antarctic 9, 106
  West Greenland 23
ice shelf lakes 48–51
ice shelf ponds 48–49, 51

ice shelves 30, 47–52
  Amery Ice Shelf 47, 133
  George VI Ice Shelf 47, 133
  Larsen Ice Shelves 47, 133
  Mackenzie Ice Shelf 83–84
  Markham Ice Shelf 49, 50–51, 133
  McMurdo Ice Shelf 47–48, 51
  Ronnie Ice Shelf 47, 133
  Ross Ice Shelf 133
  Ward Hunt Ice Shelf 47–48, 51–52, 133
ice shelves mass balance 47–48
ice worms 57
incident radiation 15
Intergovernmental Panel on Climate
    (IPCC) 9
interstitial freezing 24
iron 107
iron Fe(II) 102, 111
iron Fe(III) 104, 109, 111

Janthinobacterium 42
Janthinobacterium vividum 36
John Evans Glacier
  bacterial diversity 16–107
Jovian Moons 125–126

Kamb ice stream 106, 114
Kangerlussuaq
  community respiration 70
  photosynthetic rate 66
katabatic winds 22, 94
Kelliocottia 97
kerogens 105
krill (*Euphausia superba*) 14, 16,
    78–79, 83, 89–90

Labrador, pack ice 82
Lacrymaria 18, 97
Lac Saint-Pierre
  ice cover 98–99
  community structure 99
Lagrandian measurements 138
lake ice 2, 16–19, 92–99
lake ice annual 92–93, 96–99
lake ice break-up dates 92–93
lake ice freeze-up dates 92–93
lake ice microbial communities
    (LIMCO) 19, 95–99
lake ice perennial 92, 94–96
Lakes
  Lake Baikal 93, 99
  Beaver Lake 61, 66, 68
  Lake Bonney 95–96, 107, 109, 118
  Char Lake 61, 66
  Colour Lake 94
  Crooked Lake 6, 66
  Deep Lake 120

Lake Druzhby 6
Lake Fryxell 4, 62, 96
Gossenköllesee 18, 97
Greater Lake Priscu 107
Lake Hoare 96
Lake Mendota 93
Meretta Lake 61
Lake Miers 96
Ossian Sars 62, 66
Pendant Lake 18
Lac Saint-Pierre 98
Lake Redó 94, 96–99
Schwarzee ob Sölden 97
Lake Suwa 93
Lake Untersee 96
Lake Vida 96, 117–120
Lake Vostok 31, 103, 112–117
Lake Baikal
  ice breakup dates 93
  ice diatoms 99
Lake Bonney
  assimilation number 96
  bacterial concentration 95
  chlorophyll *a* 95
  cyanobacteria 95–96
  dissolved organic carbon 95
  perennial ice cover 95
  photosynthetic efficiency 96
  photosynthetic rate 95–96
Lake Druzhby
  viral induced bacterial
      mortality 6
Lake Fryxell
  assimilation number 96
  bacterial concentration 62
  cryptophytes 4
  lake ice community 96
  photosynthetic efficiency 96
Lake Hoare
  assimilation number 96
  ice communities 96
  photosynthetic efficiency 96
Lake Mendota
  ice breakup dates 93
Lake Miers
  assimilation number 96
  ice cover 96
  photosynthetic efficiency 96
Lake Redó
  bacterial concentration 98
  bacterial production 98
  chlorophyll *a* 97
  community structure 97
  ice cover 94, 96–98
  taxonomic diversity 98
Lake Suwa
  ice freeze-up dates 93

Lake Untersee
  laser-induced fluorescence
      emission (LIFE) 96
  microbial microcosms 96
  perennial ice cover 96
Lake Vida
  assimilation number 96
  bacterial concentration 119
  bacterial diversity 119
  chlorophyll *a* 118
  dissolved organic carbon 118
  ice cover 117–120
  photosynthetic rate 118
  photosynthetic efficiency 96
  salinity tolerances 119
Lake Vostok
  accretion ice 115–116
  bacterial diversity 115
  ecological scenarios 116–117
  hydrothermal sources 103, 113
  methanotrophy 115
  major ions 114
  subglacial lake 31, 112–117
  water chemical composition 114
landfast sea ice 73–77
Larsen Ice Shelf
  among largest 47
  catastrophic breakup 47
  exploration 133
Last Glacial Maximum (LGM) 8–10
laser-induced fluorescence emission
    (LIFE) 96, 135–137
Laurentide Ice Sheet
  Last Glacial Maximum 9
Leptolyngbya 49, 61
Leptolyngbya antarctica 51
Leptolyngbya frigida 51
light attenuation 66
lipid membranes 78
liquid chromatography mass
    spectrometry (LC-MS/MS) 141
Little Ice Age (LIA) 9–10
Longyearbreen
  community respiration 70
  photosynthetic rate 66
  sulphide oxidation 71
low temperature enzymes 40
Lyngbya 50
lysogenic cycle in viruses 5, 63, 89
lytic cycle in viruses 5, 63, 88–89

Mackenzie Ice Shelf
  bacterial concentrations 84
  bacterial morphology 83
  bacterial production 84
Magnetic Resonance Spectroscopy
    (NMR) 138

manganese Mn(II) 102
manganese Mn(IV) 109
Mantoniella 15
marginal ice zone 81
Markham Ice Shelf
　algal mat communities 49–50
　exploration 133
　surface lakes 49
　zonation of pigments 49–50
Marinobacter 84, 119
Mars 121–125
Mars Express 125
Mars Science laboratory (MSL) 125
Martian atmosphere 124–125
Martian crust 123
Martian cryospheric environment 123
Martian glaciers 123
Martian polar ice caps 122
Martian meteorite 121
Martian soil 121
Martian subglacial
　environments 123
Martian Southern Ice Cap 124
mass balance of glaciers 20
McMurdo Dry Valleys 18–19, 30, 55,
　60–62, 94–96, 106–107, 117, 139
McMurdo Ice Shelf
　ablation 47
　accumulation 47
　cyanobacteria 51
　nitrogen budget 51
　sediment 47–48
　surface ponds 51
McMurdo Sound 75, 82, 86
meiofaunal ingestion rates 91
Melosira 99
Meretta Lake
　chlorophyll *a* 61
meteoric ice melt 113
melt water drainage 138–139
melt zone 21, 29, 134
Mesodinium rubrum 15, 81, 86
Mesotaenium berggrenii 54
Mesozoic 121
metabolic rates of microbes 112
metagenomics 140–141
metanaupilius 90
methane on Mars 124–125
methane on Enceladus 127
methane on Titan 127
Methanoarchaea 7
Methanococcoides burtonii 141
methanogens 30, 111, 127–128
methanogenesis 104
Methylobacillus 115
Methylophaga 84
Methylophilus 115

microbial communities 2
　subnivial 42
microbial loop 3–6, 15–16
microorganism size categories 3
Midtre Lovénbreen
　ammonium assimilation 53
　bacterial concentrations 54,
　　56–57, 63
　bacterial production 65
　community respiration 70
　cryoconite mapping 134
　photosynthetic rate 66, 68
mixotrophy 3, 8, 16, 81, 86
mobilomes 142
modeling 139–140
Monodinium 63
moraines 28, 70–71
Mortierellales 42
moulin 25, 101
Mucorales 42
multi-year sea ice 73–77
Myoviridae 88

National Aeronautics & Space
　Administration (NASA) ICESat
　Platfrom 133
National Science Foundation Long
　Term Ecosystem Research
　Program 129
nanobacteria 121
nauplii 90–92
Navicula 48, 79
Navicula pelagica 79
nematodes 61, 90
net autotrophy 68
net heterotrophy 68
nitrate (NO_3-N) 39–40, 44–45, 71, 140
nitrate detection 137
nitrogen 3, 5, 39, 56, 64, 74, 126
nitrogen cycling 53
nitrogen fixation 18, 50–51, 56, 96
Nitzschia frigida 79
Neoproterozoic glaciations 10–12
Niwot Ride, Colorado 37, 42
Nodularia 48, 50–51
Northwest Passage 74
Nostoc 46, 48, 50, 61, 95
nutrient limitation 75–76
Nye channels 102

Oceanospirillum 84
Ochromonas 36, 38, 48
Ochromonas smithii 38
Onisimus nanseni 89–90
operational taxonomic units (OTU) 61
optical sensors 135
optimum growth temperatures 39

organic carbon 56–57, 58, 64, 65,
　102, 115, 117
organic haze on Titan 127
Oscillatoriales 61
Oscillatoria 48, 50–51, 95
osmoregulation 78
Ossian Sars
　bacterial concentrations 62
　photosynthetic rate 66
outburst floods 102
oxygen 3, 48, 104
ozone depletion 75

pack ice (sea) 73–75
Paenobacillus 111
palaeoclimate records 112
Paleozoic 121
pancake ice 74
Paraphysomonas 63
Passive Microwave
　Observations 129
Patagonian Ice Sheet
　Last Glacial Maximum 9
Patriot Hills 64, 65
Pendant Lake
　annual ice cover 18
Peptide mass fingerprinting
　(PMF) 141
percolation zone 21
perennial lake ice 17–19, 94–96
pH of snow cover 37
pH of subglacial environments 101
phage-host systems 88
Phormidium 18, 48, 50, 61, 95
Phormidium priestleyi 51
phosphorus 3, 5, 39
photoinhibition 15, 38
photolysis of methane 127
photosynthate 96
photosynthesis 3–4, 7, 15, 18, 24, 38,
　48, 50, 52, 53, 64, 66–69, 71,
　79–83, 95–96, 118, 140
　see also primary production
photosynthetically active radiation
　(PAR) 3–4, 38, 61, 66–68,
　76–77, 99
photosynthetic efficiency 68, 96
photosynthetic pigments 3
phylotypes 84
Phylum Ciliophora 7
Phylum Sarcomastigophora 7
phytoflagellates 3–4, 15–16, 30, 35,
　53, 61, 82 *see also* cryptophytes,
　*chrysophytes, dinoflagellates and
　euglenoids*
Phytomastigophora 8, 35
phytophenolic compounds 38

phytoplankton 76
Pinnularia cymatopleura 48
Planctomycetales 18
Pleurastrum 49
plumes 126
Polaribacter 84
Polarella glacialis 16
Polaromonas vacuolata 105, 106
Polyarthra 97
polycylic aromatic hydrocarbons
    (PAHs) 121
Polyketide synthases (PKSs) 78
Polykrikos 86
polymerase chain reaction (PCR) 7
polythermal glaciers 102–103
polythermal ice regime 25
polyunsaturated fatty acids
    (PUFAs) 78, 91, 142
Porosira glacialis 80
primary production 52, 64, 75–76,
    79–83, 95, 97–98, 118
    see also photosynthesis
proteomics 141
Protoperidinium 86
Protozoa 2–5, 7–8, 13
Prydz Bay 16, 75, 79, 81, 83–84, 86, 88
Pseudoalteromonas 83–84
Psychrobacter 84, 119
Psychromonas 84
psychrophilic adaptation 77
psychrophilic bacteria 30, 105
psychrophilic Cyanobacteria 48
Pyramimonas 6, 15, 81
Pyramimonas gelidicola 82
Pyramimonas tychotreta 16

Q_{10} 43
Qiyi Glacier
    cyanobacteria 56

radio-labelled carbon 115, 121
radio-labelled leucine 35, 64, 109
radio-labelled thymidine 35, 64, 109
Raman spectroscopy 137
Raphidonema nivale 35, 38–39
ratio of ice thickness to lake size 94
rDNA 115
Red River 93
red snow 2, 35
REDOX reactions 71, 103, 108, 114,
    126–128
remote sensing 21, 26, 129, 133–135
RNA tree 141
RNA world 122
Resolute Passage 83, 85, 88
respiration aerobic 116
respiration community 64, 69–71

respiration microbial 41
16S rRNA 83, 105–106, 140
Rhodoferax antarcticus 105–106
Ronnie Ice Shelf
    among largest 47
    exploration 133
Ross Sea Embayment 107
Ross Ice Shelf
    exploration 133
Rothlisberger channels 102
rotifers 4, 45, 54, 57, 78, 90–92, 97, 99
Rotmoosferner Glacier
    bacterial concentrations 63
    bacterial production 65
rotten ice 54
RUBISCO 77–78

salinity variation 14, 78
Salix polaris 45
Sample Analysis at Mars (SAM) 125
satellite derived estimates 21
satellite images 13, 47, 133
Satellite Probatoire d'Observation
    de la Terre (SPOT) 134
Saturnian moons 126–128
Saturnian E-ring 126
Saxifraga oppositifolia 45
Schwarzee ob Sölden
    ice cover 97
scuticociliates 87
Scytonema 95
scytonemin 49–50
Seas:
    Baltic Sea 84
    Barents Sea 75
    Barrow Sea 84
    Bellinghausen Sea 75, 80, 84
    Bering Sea 89
    Chukchi Sea 89
    Laptev Sea 84
    Ross Sea 75, 83, 88, 90
    Scotia Sea 75, 81
    Weddell Sea 76, 79–84, 86, 90
sea ice 2, 12–16, 73–92
sea ice algae 75–77, 79–83, 140
sea ice communities 14–16, 78–83
sea ice cover 73–75
sea ice extent 13–14
sea ice formation 73–75
SeaWifs 133
sensor technology 135–139
serpentinization reactions 127
Shannon index 110
Shewanella 83–84
Shewanella frigidimarina 88
shuga 74
Signy Island 38–39, 53

silicate weathering 103
Siphoviridae 88
slow water velocities 54
slush 53, 97
snow, activity under 41–44
snow algae 2, 35, 37–39, 134
snow bacteria 35–36, 39–41
snow biological activity 35–36
snow chemical composition 33–34
snow cover seasonal 32–33
snow cover permanent 32–33
snow cover litter laden 34
snow crystals 31, 32–33
snow depth 46
snow flakes 31–32
snow formation 31–32
snow fungi 42–44
snow metamorphism 32–33
snow melt 33, 34, 53, 101
snow molds 42–44
snow packs 32–33, 37, 44, 52
snow pack hydrology 33
snow water equivalent (SWE) 41
snow, wet 52–54
Snowball Earth 10–12
snowfall 19–21
soil carbon content 42
soil temperature 41
soil surface snow interface 42
solute in snow packs 34
South Orkney Islands 20
South Pole 19
Southern Ocean 12, 14, 16, 74–77, 83
specific growth rates of snow
    algae 39
Sphingomonas 62, 84
Spitzbergen 40
Strombidium 15, 18, 63
Stubacher Sonnblickkees
    bacterial concentrations 63
    bacterial production 65
    community respiration 70
    photosynthetic rate 66
subglacial drainage systems 26
subglacial lakes 26–27, 29, 102
subglacial tills 10
sub-ice sheet environments 138
sublimation 32
subsurface ecosystems 24–27
subsurface melting 21–22
supraglacial channels 54, 138, 139
supraglacial drainage 55, 102
supraglacial flow 101
supraglacial kames 62, 70–71
supraglacial lakes 23–24, 26–27, 30,
    55–56, 102, 111–117, 128
subglacial ocean 126

supraglacial streams 66
superimposed ice 20
surface glacier ice 20
surface melt 22, 101
sulphate (SO$_4^{2-}$) 48, 71, 104, 107, 126
sulphate reduction 108, 128
sulphides 102–103, 107–108
Svalbard 19, 23, 37, 44–45
Svalbard Bank 75, 79
sympagic crustceans 89–91
Synechoccus 95
Synthetic Aperture Radar
    (SAR) 21

tardigrades 56, 57
Tateyama Mountains (Japan)
    35–36, 40
Taylor Dome 111
Taylor Glacier
    Blood Falls 107–108
    invertebrates 57
Taylor Valley 19, 56–57, 64, 107
Temperature Gradient
    Metamorphism (TGM) (of
    snow) 32–33
temperature optima for growth
    39, 48
Terra Nova Bay 90
Terrestrial analogues 128
Thalassiosira antarctica 80
Thermocetogenium phaeum 109
Thiobacillus 106
Thiomicrospira arctica 109
Thwaites Glacier
    organic carbon and
        nutrients 129
'Tiger Stripes' 126
Titan 127–128

Trochiscia 54
turbellarians 78, 91–92
Tuva Glacier
    bacterial metabolism 71
    flow paths 53

ultra-microbacteria 65, 110
ultra violet induced
    fluorescence 135
ultra violet radiation (UV) 38, 50
unmanned airborne vehicles
    (UAVs) 134
unmanned underwater vehicles
    (UUVs) 134
Urostyla 97
Variovorax paradoxus 36
Vatnajükull ice cap 103, 128
vertical gene flow 141–142
Vestfold Hills 15, 64, 65–66, 70–71
Vestre Brøggerbreen
    community respiration 70
    photosynthetic rate 66
Viking 1976 Mission 121
viral abundance 30, 88
viral burst sizes 63, 89
viral infection 63
viral lysis 5–6
viral lysogenic cycle 63, 89
virus to bacterium ratios (VBR) 88
viruses 3–6, 63, 88–89
volcanoes 103
Volvocales 97
Voyager 2, 126

Ward Hunt Ice Shelf
    bacterial production 52
    cyanobacteria 52
    exploration 133

microbial mats 51
photosynthetic rate 51–52
remnant ice shelf 47
surface lakes 48
surface lake conductivity 52
Warm-based areas 101–102
warm based glaciers 26, 105
warm ice thermal regime 25
water-ammonium ocean on Titan 127
water equivalent (w.e.) 19, 97
water laden till 102
water residence times 103
water in subglacial habitats 25–27
weathering crust 54
Werenskiolbreen
    bacterial concentrations 71
    organic carbon 64
West Antarctic Ice Sheet
    Last Glacial Maximum 9
    subglacial sediments 106
West Greenland Ice Sheet
    supraglacial lakes 23
wet snow 52–54, 129
whaling fleet 14
Whillans Ice Steam Subglacial
    Access Research Drilling
    (WISSARD) 133
White Glacier
    desmids 61
    net mass balance 21
Worthington Glacier
    annual runoff 20

xanthophyll 15

zoochlorellae 8
Zoomastigophora 8
Zygomycota 42